36 - 44

Pitt Series in Policy and Institutional Studies

THE
ACID RAIN
CONTROVERSY

James L. Regens
Robert W. Rycroft

UNIVERSITY OF PITTSBURGH PRESS

Published by the University of Pittsburgh Press, Pittsburgh, Pa., 15260
Copyright © 1988, University of Pittsburgh Press
All rights reserved
Baker and Taylor International, London
Manufactured in the United States of America

Second printing 1989

Library of Congress Cataloging-in-Publication Data

Regens, James L.
  The acid rain controversy.

  (Pitt series in policy and institutional studies)
  Includes index.
  1. Acid rain—Environmental aspects—United States.
2. Environmental policy—United States. I. Rycroft,
Robert W., 1945–   . II. Title. III. Series.
TD196.A25R44 1988   363.7'386   87-35769
ISBN 0-8229-3582-1
ISBN 0-8229-5404-4 (pbk.)

TO

LESLIE AND MARILYN

CONTENTS

TABLES		ix
FIGURES		xi
ABBREVIATIONS		xiii
ACKNOWLEDGMENTS		xvii
1	The Emergence of the Acid Rain Controversy	3
2	The Science of Acid Rain: An Evolving Consensus?	35
3	Acid Rain Control Technologies and Mitigation Strategies	59
4	The Economic Dimension: Benefits, Costs, and Financing Options	85
5	Acid Rain as a Political Problem	113
6	Prospects for Policymaking	135
APPENDICES		169
NOTES		195
REFERENCES		201
INDEX		219

TABLES

1. Chronology of Major Federal Legislation to Control Air Pollution — 8
2. National Ambient Air Quality Standards, Major Effects, and Emissions Sources for Criteria Pollutants — 14
3. Status of Toxic Air Pollutant Evaluation and Control Program, 1985 — 16
4. National Expenditures for Pollution Abatement and Control, 1972–1983 — 23
5. Relative Contribution of Natural and Anthropogenic Sources to Total Sulfur and Nitrogen Oxides Emissions in the United States — 37
6. Estimated Number of Tall Stacks Constructed in the United States, 1970–1979 — 47
7. Opposing Interpretations of Existing Scientific Knowledge as Rationales for Acid Rain Policy — 52
8. NAPAP Funding of Agencies, FY 1982–FY 1987 — 55
9. NAPAP Funding of Task Groups, FY 1982–FY 1987 — 56
10. Available, Potential, and Possible Technologies for Controlling Acid Deposition — 60
11. Existing and Projected Flue Gas Desulfurization Systems in U.S. Utilities — 65
12. Annual Maximum Economic Losses Attributable to Acid Deposition in the Eastern Third of the United States — 86
13. Scientific Data Needed for Estimating Economic Benefits of Controlling Acid Deposition — 87
14. Incremental Costs of Strategies for Reducing SO_2 Emissions — 92

15. Electricity Rate Increases for a 10 Million Ton Reduction in SO_2 Emissions — 100
16. Distributional Impact of Selected Options for Financing Reductions in Acid Deposition — 105
17. Grants Under the EPA State Acid Rain Program — 145
18. H.R. 4567 Sponsorship by State — 155

FIGURES

1. The pH Scale   36
2. Schematic View of the Acid Deposition Problem   39
3. Annual Mean Value of pH in Precipitation in the United States and Canada, 1980   44
4. Sulfur Dioxide and Nitrogen Oxides Emissions in the United States and Canada, 1980   46
5. Potential Direct and Indirect Effects of Acid Deposition and/or Sulfur and Nitrogen Oxide Emissions   49
6. Process Flow for a Typical Physical Coal Cleaning Plant   62
7. Limestone Wet Scrubbing FGD Process   66
8. Saarberg-Hölter Dry FGD Process   68
9. Wellman-Lord Regenerable FGD Process   70
10. Typical Low NO_x Burner   72
11. U.S. Sulfur Dioxide and Nitrogen Oxides Emissions Trends, 1900–2030   79
12. Quantities and Sulfur Content of Coals Produced for Electric Utilities by State, 1980   93
13. The Policy Dilemma   163

ABBREVIATIONS

AFBC	atmospheric fluidized bed combustion
ANC	acid neutralizing capacity
BACT	best available control technology
Btu	British thermal unit
Ca	calcium
CAA	Clean Air Act
CCC	chemical coal cleaning
CEM	continuous emissions monitoring
CEQ	Council on Environmental Quality
Cl	chlorine
CWIP	construction-work-in-progress
DDT	dichloro-diphenyl trichloroethane
DHEW	U.S. Department of Health, Education and Welfare
DOE	U.S. Department of Energy
DOI	U.S. Department of the Interior
EACN	European Air Chemistry Network
EBI	electron beam irradiation
ECE	United Nations Economic Commission for Europe
EDF	Environmental Defense Fund
EEC	European Economic Community
EEI	Edison Electric Institute
EPA	U.S. Environmental Protection Agency
EPRI	Electric Power Research Institute
FGD	flue gas desulfurization
FGT	flue gas treatment
FY	fiscal year
H^+	hydrogen
HC	hydrocarbons

HCO_3^-	bicarbonate
IGCC	integrated gasification combined-cycle
K^+	potassium
kwh	kilowatt hour
LEA	low excess air
LIMB	limestone injection multistage burner
Mg	magnesium
MHD	magnetohydrodynamics
MW	megawatt
MOI	U.S.-Canada Memorandum of Intent on Transboundary Air Pollution
Na	sodium
NAAQS	national ambient air quality standards
NAPAP	National Acid Precipitation Assessment Program
NAS	National Academy of Sciences
NCAQ	National Commission on Air Quality
NESHAPS	national emissions standards for hazardous air pollutants
NGA	National Governors' Association
NH_4^+	ammonium
NO_x	nitrogen oxides
NO	nitric oxide
NO_2	nitrogen dioxide
NO_3^-	nitrate
NOAA	National Oceanic and Atmospheric Administration
NRDC	Natural Resources Defense Council
NSPS	new source performance standards
NWF	National Wildlife Federation
O_3	ozone
OECD	Organization for Economic Cooperation and Development
O&M	operating and maintenance
OSTP	Office of Science and Technology Policy
PCC	physical coal cleaning
PFBC	pressurized fluidized bed combustion
pH	potential hydrogen
PSD	prevention of significant deterioration
PUC	public utility commission

R&D	research and development
SAB	Science Advisory Board
SCR	selective catalytic reduction
SIP	state implementation plan
SNR	selective noncatalytic reduction
SNSF	Norwegian Interdisciplinary Research Project
SO_2	sulfur dioxide
SO_3	sulfur trioxide
$SO_4^=$	sulfate
STAPPA	State and Territorial Air Pollution Program Administrators
STAR	State Acid Rain Program
Tg	Terragram
USDA	U.S. Department of Agriculture
VOC	volatile organic compound
μeq/l	microequivalents per liter

ACKNOWLEDGMENTS

ARISTOLE's *Metaphysics* opens with the observation: "All men by nature desire to know." The initial impetus for writing this book reflects such an orientation. Our analysis grew out of a series of discussions between the two of us about various aspects of the United States government's response to the acid rain problem while one of us (Larry Regens) was with the Environmental Protection Agency, from August 1980 to September 1983. Several questions dominated those conversations, and this book was written in an attempt to grapple with them. How did acid rain become a public policy issue? Given the existing scientific knowledge, is a consensus emerging as to its causes, its effects, and the severity of those impacts? What control technologies and mitigation strategies are available for preventing adverse effects? How much do we know about estimating the economic costs and benefits of acid rain and assessing alternatives for financing emissions control programs? How is acid rain a political problem? What are the prospects for policymaking? This study strives to provide answers to those questions.

During that 1980–1983 period, Larry served as EPA joint chair of the Interagency Task Force on Acid Precipitation (1981–1982) and was involved actively in the formulation and implementation of the large-scale, federal research effort addressing the acid rain question. His participation in the negotiations on transboundary air pollution between the United States and Canada, including the technical work groups supporting that effort, offered an opportunity to gain a deeper appreciation of the dynamics of interaction among diverse interest groups, the bureaucracy, Congress, and the media in this policy arena. Similarly, Larry's service as chair of the Energy and Environment Group of the Organization for Economic Cooperation and Development from 1981 to 1983, and as a U.S. delegate to numerous meetings of the United Nations Economic Commission for Europe, provided another dimension—that of the international context. As a result, we believe that our book brings a special blend of scholarly analysis and personal insights to bear on the acid rain controversy.

We are grateful to many colleagues for their stimulating discussions. A number of people with government, environmental groups, trade associations, and the research community gave generously of their time and expertise in order to provide valuable substantive advice and encouragement in the completion of this book. Some of them may not have realized the major effect that a seemingly casual conversation had on our subsequent thinking. In particular, we would like to thank John Bachmann, Rob Brenner, Thomas H. Brand, Jr., Robert W. Brocksen, Ellis Cowling, Thomas D. Crocker, Robert M. Friedman, David G. Hawkins, George Jordy, Lester Machta, Michael Oppenheimer, George Rejohn, Courtney Riordan, Hans Martin Seip, and Michael Tinkleman. Also, an anonymous reviewer for the University of Pittsburgh Press offered helpful criticism on an earlier version of the manuscript. Of course, while we thank them, we acknowledge that none of these people should be held responsible for the contents of this book.

The staff of the University of Pittsburgh Press, especially Frederick A. Hetzel and Bert A. Rockman, exceeded our most optimistic expectations from beginning to end. Closer to home, the typing of this manuscript was no easy task. We would like to thank Alisa Chandler, Paula Johnson, and Amy Williams, each of whom is at the University of Georgia's Institute of Natural Resources, for their patient and superb efforts.

Also due special thanks are Larry's sons, Craig and Kent, who provided timely interruptions and reminders of the need to maintain perspective throughout this endeavor. Lastly, we appreciate the invaluable contributions and gentle pleas to complete the manuscript from our wives, Leslie and Marilyn, to whom this book is dedicated. Without their persistence, this volume could not as readily have gone beyond the stage of spirited discussion between its authors.

JAMES L. REGENS
ROBERT W. RYCROFT

THE ACID RAIN CONTROVERSY

1
THE EMERGENCE OF THE ACID RAIN CONTROVERSY

The acid rain problem vividly demonstrates the need for new approaches to environmental management. Rather than an aberration, acid rain is probably the prototype of the new environmental problems that increasingly will confront the nation.
—*Conservation Foundation (1984)*

OVER THE PAST decade, the phenomenon of acid deposition, commonly referred to as "acid rain," has been the subject of growing scientific research as well as widespread media coverage.[1] The acidification of land and water has come to be described in drastic terms, ranging from "perhaps the worst environmental threat ever to hit us," to "an ongoing environmental catastrophe," to "our biggest environmental problem now and for the future" (Swedish Ministry of Agriculture 1982, 8). Such words reflect the depth of concern which has transformed acid rain from an esoteric topic of scientific research in certain specialized fields of ecology and atmospheric chemistry into a household word.

In the process, acid rain has emerged as one of the most controversial environmental problems of the 1980s. The burning of fossil fuels has released vast quantities of pollutants into the atmosphere. Once airborne, these pollutants often form acidic compounds that frequently travel long distances before returning to the earth's surface in fog, mist, rain, or snow, as well as in dry form. While the substances that produce acid rain also occur naturally, it is the man-made contributions that have transformed acid rain into an environmental problem of potentially global proportions. For example, many have expressed concern about the possible short- and long-term consequences of the deposition of sulfuric and nitric acids on aquatic ecosystems, forests, crops, and manmade structures or cultural artifacts. This has been the case particularly in eastern North America and Europe, although concern is beginning to emerge elsewhere (see Ambe and Nishikawa 1986).

Such attention has fostered a corresponding and dramatic increase in the general public's awareness of acid rain's existence. For example, in 1980, few Americans identified acid rain as a policy problem. On the other hand, as early as 1980, a Gallup poll found that two-thirds of Canadian adults were aware of acid rain as an environmental concern (Roeder and Johnson 1985). Only three years later, however, a Harris poll found that a rapid expansion of awareness comparable to Canadian familiarity with the issue had occurred in the United States. Sixty-three percent of those questioned were aware of acid rain and approximately two-thirds of the U.S. public favored stricter controls on sulfur dioxide (SO_2) emissions, one cause of acid rain.

Almost at the same time, acrimonious debate about the extent to which acid deposition constitutes an environmental risk emerged within both the scientific community and the broader political arena. Disagreement about the merits of various approaches to address the acid rain phenomenon has become a focal point for the ongoing debate surrounding the reauthorization of the Clean Air Act (CAA) by the United States Congress. It also has become a serious foreign policy issue clouding bilateral relations between the United States and Canada, as well as creating transnational conflict within Europe (see Schmandt and Roderick 1985; Carroll 1983). These tensions have generated growing pressure to address the problem, and the highest levels of government have been forced to make room on the political agenda for acid rain concerns. As a consequence, acid deposition has been transformed from a relatively obscure area of scientific inquiry into one of the paramount environmental issues of the 1980s.

Politics, science, and economics dominate the evolution of the acid rain issue. Much of the politics, especially regional and inter-group conflict, inherent in the current acid rain controversy was predictable, given the history of air pollution control efforts in the United States. That is, the conflict over whether to impose further emissions reductions as an acid rain control strategy is to some extent a surrogate for broader debates about efforts to improve ambient air quality. Those debates about the nation's regulatory framework for air pollution have been raging for the better part of two decades.

While the political dimension has proven to be somewhat predictable, major aspects of the scientific and economic dimensions of the acid rain conflict were more difficult to anticipate because of several factors. First, from an ecological standpoint, the acid rain problem not only covers long periods of time but also affects air, water, soil, and biota. Those impacts transcend political boundaries. Second, the economics of acid rain involve a known high level of reduction costs coupled with a questionable level of benefits. Such an imbalance makes it extremely difficult to establish equitable costs and benefits of proposed regulatory actions. Finally, the acid rain problem involves substantial policy uncertainty and complexity because it involves several scientific disciplines. In other words, the acid deposition problem serves as a striking paradigm for a great many emerging environmental problems. Its scope is international; its causes are scattered in a variety of countries. Similarly, its consequences extend beyond existing political boundaries so that a successful resolution will require cooperative action. Taken together, these factors have created a classic environmental policy controversy (Regens and Rycroft 1985; 1986).

THE POLITICAL CONTEXT OF ACID RAIN: AIR QUALITY

For at least the past two millennia, air pollution has been looked upon as a nuisance. As early as A.D. 61, the philosopher Seneca noted Rome's polluted vistas. Almost a thousand years later, the pollution associated with wood burning at Tutbury Castle in Nottingham was considered unendurable by Eleanor of Aquitaine, the wife of King Henry II of England, forcing her to move. Moreover, starting as early as 1273, a series of royal decrees were issued barring the combustion of coal in London in a futile attempt to address that city's burgeoning air quality problem. Such illustrations are more than curious anecdotes;

they underscore the enduring nature of air pollution as a public concern.

Some five hundred years later, in the eighteenth and nineteenth centuries, modern pressures for air pollution control began with the onset of the Industrial Revolution. First in England and later in Western and Central Europe, as well as in the United States, the advent of the widespread use of steam power for commercial purposes was coupled with growing local air quality problems. As early as 1819, the British Parliament was undertaking studies of pollution abatement measures. The first technological fixes, such as stokers and flue gas scrubbers, became attractive during the nineteenth century.

In the United States, rapid technological change between 1900 and 1925 was especially important in bringing both stationary sources, such as factories and power generating stations, and mobile sources of air pollution, such as rail and automobile transportation, to the attention of the general public. Almost forty years ago, many contemporary air quality problems and their proposed solutions had emerged. For example, localized health crises due to increased air pollution had been experienced in Los Angeles, California, and Donora, Pennsylvania.[2] The first professional meeting of air quality experts had taken place, and large-scale research on air pollution had begun. Moreover, initial efforts to regulate air quality were under way, although only in California. Nonetheless, while the increase in smoke and ash was considered to be a human health concern, as well as a public nuisance, policy responses clearly remained circumscribed. Until the mid-twentieth century, government regulations, scientific understanding of the problem, and public attitudes viewed air pollution control as an exclusively municipal responsibility in the United States (Regens 1985a).

It was not until a great fog blanketed London from December 5 to December 8, 1952, that the awesome potential for adverse health effects associated with air pollution became more apparent. Ten days later it was learned that the total number of deaths in the Greater London area for that period exceeded the average by 4,000, with almost all of the unexpected deaths attributed to bronchitis, emphysema, or cardiovascular problems (see Lave and Seskin 1977). The 1952 disaster in London, coupled with serious air pollution problems in a number of other industrialized countries including the United States, accelerated the establishment of research centers and paved the way for more direct action by the U.S. government.

Table 1 provides a chronology of the federal legislation and the evolving national and state roles dealing with air pollution problems in the United States. The first air pollution legislation adopted at the federal level in the United States was passed in 1955. At that time, the federal government's effort was limited to research and development (R&D), training, and technical aid to state governments. These services were provided by the Department of Health, Education and Welfare's (DHEW) Public Health Service (Stern et al. 1984, 3–17). Thus, while air pollution physically transcends political boundaries, it was considered from a policy standpoint to be a purely local problem with abatement activities a state and local function. As a result, six years after the initial federal grant-in-aid program was started, only seventeen states operated air pollution control programs with annual budgets in excess of $5,000 (U.S. Department of Health, Education and Welfare 1969a, 11).

Eight years later, the Clean Air Act (CAA) of 1963 was passed. While continuing the existing emphasis on federal support of scientific and technical advice to the states, the CAA also expanded somewhat the role of the federal government as a facilitator of intermunicipal and interstate air quality control efforts. The 1963 act expanded the state grants-in-aid program too. However, this early effort to control pollution had a narrow focus. It reflected a high level of concern for motor vehicle emissions but failed to provide for any programs dealing with other sources. Moreover, although the 1963 act also established a procedure for federal prosecution of interstate air pollution cases by the Department of Justice, the procedure in practice proved to be a largely ineffective mechanism for addressing interstate air pollution problems.

Two years later, Congress passed the Motor Vehicle Pollution Control Act of 1965. That act authorized DHEW to set federal standards for new automobile emissions. However, neither the 1963 nor the 1965 legislation provided for the establishment of national emission standards for stationary sources such as utility and industrial polluters at fixed sites.

The nation's current approach to air pollution control was

TABLE 1
Chronology of Major Federal Legislation to Control Air Pollution

	Federal Role	State Role
Air Pollution Control Act of 1955	Provide research, technical, and financial assistance to states	All responsibility for control
Clean Air Act of 1963	Mediate among states, if requested	Form regional commissions
Air Quality Act of 1967	Create air quality control regions; establish criteria for health protection; recommend control techniques; set national emissions standards for vehicles	Must adopt ambient air quality standards subject to federal review and approval
Clean Air Act Amendments of 1970	Set national primary and secondary air quality standards; review and approve state implementation plans; assess hazards from additional named pollutants; set national emissions standards for stationary sources; set statutory reductions and timetable for vehicle emissions; regulate fuels, fuel additives, aircraft emissions, noise	Design and enforce state implementation plans, if approved by EPA; right to impose more stringent standards
Clean Air Act Amendments of 1977	Classification of air quality control regions as attainment or nonattainment; program for prevention of significant deterioration; special treatment for eastern coal; new source performance standards and hazardous pollutant sections strengthened; motor vehicle emissions standards tightened further	Modification of state implementation plans for nonattainment areas to avoid major sanctions, cost-benefit analysis and offset policy for new sources
Acid Precipitation Act of 1980	Creation of Interagency Task Force on Acid Precipitation; authorized ten-year National Acid Precipitation Assessment Program	No specific role established

established by the Air Quality Act of 1967. Unlike previous acts, it created the first minimum national standards for overall air quality. That legislation was a major step toward expanding the federal government's role and it reflected growing congressional dissatisfaction with the CAA's reliance on state initiatives. In many respects, the Air Quality Act "was a prelude to the legislation of the 1970s" (Crandall 1983a, 7). It authorized the federal government to designate air quality control regions, recommend specific control technologies, and conduct expanded research and development. But the states retained standard-setting and enforcement responsibility. Thus, while the 1960s witnessed a gradual expansion of the federal government's role, control of air pollution was perceived to be primarily the responsibility of state and local governments (U.S. Department of Health, Education and Welfare 1969b).

Yet continued reliance on the states to take the lead in establishing and enforcing air quality standards did not appear to enhance significantly the nation's air quality. Instead, public awareness of a deteriorating environment, including declining air quality, became more pervasive throughout the 1960s. This public opinion fostered increasing support for the federal government to assume a leadership role in air pollution control (see Crandall 1983a; Friedlaender 1978; Jones 1975).

The pattern culminated in the adoption of a variety of symbolic and substantive changes in the nation's environmental management system. Starting with President Nixon's signing of the National Environmental Policy Act on January 1—his first official action of 1970—these changes signaled the dawn of what was then commonly referred to as the "environmental decade." Shortly thereafter, the Nixon administration announced a major environmental initiative in a presidential message to Congress. By the end of 1970, the Environmental Protection Agency (EPA) was established by consolidating antipollution programs that previously had been scattered throughout the federal government, and Congress adopted a series of far-reaching amendments to the existing air pollution control legislation. The CAA amendments of 1970 completed the cycle of shifting authority over the standards-setting process from local to state to federal government.

Three key features of the 1970 legislation gave the federal government additional power to deal with ambient air quality problems. First, the 1970 CAA amendments authorized the federal government to set uniform national ambient air quality standards (NAAQS) for certain pollutants. Second, the federal government was required to establish uniform emission standards for new sources of pollution, commonly termed new source performance standards (NSPS). Third, the states were required to formulate state implementation plans (SIPs) to attain those standards. As a result, for the first time the federal government assumed the responsibility of regulating emissions for new industrial sources under the NSPS formulated by EPA in 1971, as well as air pollution from motor vehicles. In addition, EPA was given the responsibility to oversee state controls on emissions from existing sources of air pollution.

In 1977 Congress expanded the CAA with additional amendments. Although compliance deadlines were extended for urban areas with severe mobile source pollution and for selected industrial sectors, the 1977 amendments, when coupled with the 1970 provisions, made the CAA even more far-reaching in its regulatory purview. For example, the 1977 amendments added three provisions to regulate interstate and international air pollution effects through the existing SIP mechanism. While section 126 permits states or their subdivisions to seek relief from interstate pollution, its language is relatively vague and relies on section 110(a)(2)(E) provisions to establish the quantity of pollution subject to controls. However, section 110(a)(2)(E) applies only to emissions from one state that cause concentrations of a pollutant in another state to exceed NAAQS and not to those pollutants for which no ambient standard exists. Section 115, which provides administrative procedures for addressing international transboundary air pollution, is not restricted to criteria pollutants, but the section gives no guidance on what levels and kinds of air pollutants are prohibited, how to allocate control responsibilities to the states, or how to revise SIPs to control pollutants for which NAAQS do not exist.

The 1977 legislation also established a program for the prevention of significant deterioration (PSD) of air quality in areas exceeding NAAQS; required new sources in nonattainment areas to "offset" their emissions by reducing those of existing sources; required EPA to identify pristine areas such as national parks which might be damaged by air pollution and to establish a program to protect those integral vistas, provided that the use of locally mined coal might be required by a governor to prevent severe economic disruption or unemployment in a particular state; and required EPA to revise the NSPS for each generic industrial source of pollution. For fossil fuel–fired

The Emergence of the Controversy 11

boilers used by electric utilities or other industries, EPA was required to specify a minimum percentage reduction in sulfur dioxide emissions based on the use of the best available control technology (BACT) for continuous emission control. Those changes in new source performance standards were incorporated into the revised NSPS adopted by EPA in 1979. The 1979 NSPS have the effect of mandating that flue gas desulfurization (FGD) systems, commonly called scrubbers, be used on all new coal-fired units, regardless of the sulfur content of the coal burned in those units.

Although scheduled for reauthorization in September 1981, the CAA instead has been the subject of intense conflict ever since that time between environmental groups and industrial interests. For example, both sides have had legislation introduced in the Senate and House of Representatives supporting their respective positions. But no substantive changes in the CAA have been enacted over the intervening years. Instead, funding for continued enforcement of the current provisions of the law has been provided by appropriations resolutions, while the debate continues in Congress over how best to address the nation's air pollution problems, including acid rain.

Although the objectives of the CAA and its amendments are relatively straightforward, the air pollution control regulations as developed by EPA are perhaps the most complex in the entire federal government. At the most general level, the national air pollution management system consists of five stages: the goal-setting, criteria-development, air quality standard–setting, emissions standard–setting, and enforcement stages (Rosenbaum 1985, 112–17; Berry 1984). Goal-setting involves a determination of the overall targets for abatement. The criteria development stage synthesizes scientific studies and technical data that indicate the relationships between various pollutants and averse effects on human health or the environment. Air quality standards establish the maximum levels of specific pollutants that will be tolerated in ambient air based on the scientific information provided by the criteria document. Emission standards then delineate the permissible levels for individual pollutants that can be released into the atmosphere from significant sources. And the enforcement stage tries to ensure that pollution standards are achieved, through a range of compliance mechanisms and sanctions. There have been important policy conflicts at each of these stages of air quality control.

THE AIR QUALITY MANAGEMENT SYSTEM

The major problems with air quality goal-setting have had to do with the paucity of empirical data upon which most of them are based. The difficulties of linking lofty goals to physics, chemistry, public health, engineering, economics, and any number of other sources of data has often returned to haunt EPA in the form of conflicts regarding the criteria and standard-setting stages. Lave and Omenn (1981, 45) describe how our understanding of this linkage has changed over time: "Standards have been established for seven pollutants: particulate matter, sulfur dioxide, carbon monoxide, nitrogen dioxide, photochemical oxidants (ozone), hydrocarbons, and lead. In three of these cases, the wrong pollutant is being regulated; fine respirable particles and acid sulfates should replace total suspended particulate matter and sulfur dioxide, and the hydrocarbons standard should be scrapped altogether." In short, there are substantial doubts as to whether several of the standards now in place are scientifically defensible. In fact, the ambient standard established for hydrocarbons was rescinded by EPA in January 1983. Originally the standard was based on studies providing evidence of adverse health or welfare effects associated with ozone (O_3) formed by reactive hydrocarbon (HC) as a class. Individual HC species currently are being evaluated by EPA for regulation as possible hazardous air pollutants. In the interim, hydrocarbons are controlled through the attainment of the ambient standard for ozone (see Berry 1984).

But other than eliminating the hydrocarbon rules, achieving flexibility in standard-setting has been next to impossible. This problem is made worse when one considers the range and complexity of the criteria and standards that have been developed. For purposes of regulation, the air quality criteria established by EPA consist of ambient standards that measure the quality of the atmosphere, emission standards measuring the quantities of pollutants discharged from facilities, and standards for hazardous pollutants found to present a risk of irreversible effects or incapacitating reversible effects. Federal ambient air quality regulations establish both "primary" and "secondary" standards. The primary standards for what commonly are called criteria pollutants are designed to set the maximum level of pollutant concentrations that is not harmful to public health based on a review of existing scientific information. The secondary standards for those criteria pollutants establish maximum permissible concentration levels to protect the public welfare (see table 2). Emission performance standards are divided into rules for "old sources" and "new sources." Using those categories, NSPS set allowable emission rates for certain pollutants for each specific type of facility and require the application of BACT, regardless of fuel quality. Adding to the complexity of the current air quality management system are the PSD and offset rules outlined above. Finally, there are separate standards involving the development of national emissions standards for hazardous air pollutants (NESHAPS) under section 112 of the Clean Air Act. Such compounds as asbestos, beryllium, vinyl chloride, and mercury represent potential hazardous air pollutants. To date, EPA has listed only eight compounds that cause a significant risk to the public, and progress in the promulgation of these rules under the NESHAPS program has proceeded at a snail's pace. As of early 1987, emissions standards had been established for six toxic air pollutants—mercury, beryllium, asbestos, vinyl chloride, benzene, and radionuclides—and proposals were pending for inorganic arsenic and coke ovens (see table 3).

Enforcing this diverse set of procedures is the responsibility of the states, through the SIP mechanism. Most of the policy controversy quite naturally has focused on this set of activities. Two issues have dominated. The first involves federal-state cooperation for developing and implementing SIPS. The second concerns whether obstacles to economic growth are presented by NSPS, BACT, and PSD requirements.

Intergovernmental relationships in the area of state implementation plans have suffered because many state officials have viewed EPA's oversight of the air quality programs as overly detailed, the source of substantial bureaucratic delay, and not flexible enough to take into account regional, state, or local interests. In short, states would like more discretionary authority over the implementation of clean air provisions. On

TABLE 2
National Ambient Air Quality Standards, Major Effects, and Emission Sources for Criteria Pollutants

Criteria Pollutant[a]	Primary		Secondary	
	Averaging Time	Concentration	Averaging Time	Concentration
Sulfur oxides (SO_2)	Annual arithmetic mean	(0.03 ppm) 80 µg/m^3	3-hour	(0.50 ppm) 1,300 µg
	24-hour	(0.14 ppm) 365 µg/m^3		
Nitrogen oxides (NO_2)	Annual arithmetic mean	(0.053 ppm) 100 µg/m^3	Same as primary	
Ozone (O_3)	Maximum daily 1-hour mean	0.12 ppm (235 µg/m^3)	Same as primary	
Particulate Matter (PM)	Annual geometric	75 µg/m^3	Annual geometric[b]	60 µg/m
	24-hour	260 µg/m^3	24-hour	150 µg/m
Carbon Monoxide (CO)	8-hour	(9 ppm) 10 mg/m^3	Same as primary	
	1-hour	(35 ppm) 40 mg/m^3	Same as primary	
Lead (Pb)	Maximum Quarterly	1.5 µg/m^3	Same as primary	

Sources: Adapted from Council on Environmental Quality (1985); Berry (1984); Lave and Omenn (19

a. The standards are categorized for long- and short-term exposure. Long-term standards specify annual or quarterly mean that may not be exceeded; short-term standards specify upper limit values for 3-, 8-, or 24-hour averages. The short-term standards are not to be exceeded more than once per year. example, the ozone standard requires that the expected number of days per calendar year with d maximum hourly concentrations exceeding 0.12 parts per million (ppm) be less than or equal to one.

b. This annual geometric mean is a guide used in assessing implementation plans to achieve the 24- standards of 150 µg/m^3.

Health Effects	Welfare Effects	Major Sources
...vates symptoms of heart and ...isease; increases incidence of respiratory diseases including ...s and colds, asthma, bronchi-...d emphysema	Is toxic to plants; can destroy paint pigments, erode statues, corrode metals, harm textiles; impairs visibility; precursor to acid rain	Electricity-generating stations, smelters, petroleum refineries, industrial boilers
...ncrease susceptibility to viral ...ions such as influenza, irritate ...ngs, and cause bronchitis	Is toxic to vegetation; causes brown discoloration of the atmosphere; precursor to acid rain	Electric utility boilers and motor vehicles
...tes mucous membranes of respiratory system, causing coughing, ...red lung function, reduced re-...ce to colds; can aggravate ...a, bronchitis, emphysema	Corrodes materials such as rubber and paint; can injure crops, trees, shrubs	Formed by chemical reactions in the atmosphere from two other airborne pollutants: nitrogen oxides and hydrocarbons
...rry heavy and cancer-...g organic compounds deep ...e lung; with sulfur dioxide, ...crease incidence and severity ...iratory diseases	Obscures visibility; causes dirty materials and buildings; corrodes metals	Industrial processes and combustion; about 7 percent from natural, largely uncontrollable, sources (windblown dust, forest fires, volcanoes)
...eres with blood's ability to ab-...xygen, thus impairing percep-...d thinking, slowing reflexes ...using drowsiness, uncon-...ness, and death. Long term ex-... suspected of aggravating ar-...lerosis and vascular disease		Motor vehicles
...s blood-forming, reproductive, ...s, and kidney systems; can ...ulate in bone and other tis-...osing a health hazard even ...xposure has ended. Children ...rticularly susceptible; behav-...bnormalities including ...ctivity and decreased learn-...lity have recently been dem-...ed		Motor vehicle exhaust; lead smelting and processing plants

TABLE 3
Status of Toxic Air Pollutant Evaluation and Control Program, 1985

Preliminary Health Screening	Detailed Assessment[a]	Science Advisory Board Review[b]	Regulatory Options Analysis[d]	Regulatory Decisions	NESHAPS Proposed	NESHAPS Promulgated
Copper	1,3 Butadiene	Chloroform	Acrylonitrile	Toluene (N)	Benzene	Mercury
Phenol	Dibenzofurans	Dioxins	Methyl chloroform	Beryllium (L)	Arsenic	Beryllium
Propylene		Nickel	Freon 113	Asbestos (L)	Arsenic (L)	Asbestos
Propylene oxide		Beryllium[c]	Carbon tetrachloride	Vinyl chloride (L)		Vinyl chloride
Acetaldehyde			Methylene chloride	Coke oven emissions (L)		Benzene
Acrolein			Manganese	Benzene (L)		Radionuclides
			Hexachlorocyclopentadiene			
Hydrogen sulfide			Gasoline vapors	Radionuclides (L)		
Chloride and HCl			Chlorobenzenes	POM (N)		
Ammonia			Epichlorohydrin	Mercury (L)		
Zinc oxide			Asbestos[d]			
Styrene			Vinylidene chloride			
			Ethylene dichloride			
			Chromium			
			Perchloroethylene			
			Trichloroethylene			
			Cadmium			
			Ethylene oxide			
			Chloroprene			
			Phosgene			

Source: Council on Environmental Quality (1985).
Note: L = listed under Section 112; N = decision not to regulate.
a. Health and exposure assessment. Not yet submitted to EPA Science Advisory Board (SAB).
b. Submitted to SAB. Recommendations not yet received.
c. Reassessment of original health effects information.

the other hand, states have been extremely nervous about recent "new federalism" initiatives designed to provide them with these additional environmental protection responsibilities, while reducing concomitant federal funding and information capabilities (McCurdy 1986).

The combination of NSPS, BACT, and PSD requirements is perhaps the most controversial aspect of federal air quality regulation. The new source performance standards are widely seen as limiting efficient economic choices, but by far the most explosive policy issue was generated when the BACT requirement for sulfur dioxide removal was interpreted to mean a blanket mandate for the use of FGD technology for all major new coal-combustion facilities, regardless of the grade of coal used. This policy ruled out supplementary or intermittent air pollution control strategies and added significantly to the costs of power generated by coal-fired plants. In addition, it worked to the decided disadvantage of low-sulfur coal producers and users in western states and provided a dramatic boost to eastern, high-sulfur interests (Ackerman and Hassler 1981).

The PSD requirement, which had been viewed by many parties-at-interest as potentially the most volatile air pollution regulation, has had less impact (see National Research Council 1981). A stagnant economy from the late 1970s through the early 1980s must receive much of the credit for reducing conflict about how much degradation, if any, should be allowed in the various air classification areas established under the terms of the 1977 legislation. The development of three classes of air quality protection, a compromise between environmentalists concerned about the deterioration of pristine regions and industries desiring rules that would allow development in areas with high air quality, made neither side happy and is still resented by some parties-at-interest in the West as an obstacle to increased development. This is the case especially when the PSD barrier is made even larger by the inclusion of visibility standards and is accompanied by EPA's offset rule, which allows development in areas already violating clean air standards (Devine et al. 1981, 127–30).

Cutting across these issues is a central theme of the federal air quality effort. In the more than fifteen years since the implementation of the first meaningful legislation, half a dozen major programs have been developed, and some air quality improvements have been recorded. But it has taken far longer than the time allotted originally. Costs have been dramatically higher than anyone anticipated, and the political conflicts have taken a heavy toll on the ability to generate consensus among the ever increasing range of participants in air pollution policy. These three key points—delays in the application of clean air rules, the economic burden inflicted on particular interests and the society as a whole by compliance with air pollution control legislation, and the divisive political consequences of the air quality debate—deserve closer attention.

The evolution of U.S. air pollution control policy has been characterized by an optimistic assessment of the ability of major institutions, legislative as well as bureaucratic, to attain mandated goals in a timely fashion. Two dimensions of the air quality experience have been especially troublesome in this respect: the action-forcing and technology-forcing aspects of the management system. First, the CAA requires EPA to undertake standard-setting and enforcement activities to reduce air pollution according to a congressional time schedule. Second, it mandates the development of necessary innovations in control technology to meet the deadlines.

As originally designed, forcing action and technology were viewed as strategies to reduce agency latitude, rationalize the regulatory process, and rapidly achieve dramatic improvements in air quality. Domination of the agency by special interests and bureaucratic sloth in formulating and implementing regulations, the presumed barriers to those goals, were to be avoided by detailed congressional intervention into and monitoring of the air quality control process. Building a legislative coalition to support adoption of this reform depended upon assumptions about the following: injection of the legislative branch into the routine affairs of regulatory agencies; the inadequacy of traditional pork-barrel or logrolling and the disjointed incrementalism approach to bureaucratic policy choices; and how state government inertia had created or at least exacerbated a perceived environmental crisis with the accompanying political need to do something in the face of recognized uncertainties.

Unfortunately, the attempts to rationalize the process and get quick fixes in air quality never achieved these goals, if they

The Emergence of the Controversy 19

were ever actually attainable. The CAA amendments were intended to be a dramatic departure from previous environmental protection initiatives, both in terms of their explicit standards and their strict timetables. But the reality of air quality policy is outlined by Melnick (1984, 127) in stark terms:

The reality of pollution control is that regulators and polluters bargain over what controls are "reasonably available," with each side keeping in mind the extent of its political support. If application of these "reasonably available" controls fails to result in attainment of national standards by statutory deadlines, then Congress either passes new deadlines (as it did for the steel industry in special legislation passed in 1982) or it lets the deadlines quietly slip away.

In other words, despite the CAA's action-forcing and technology-forcing intent as reflected in its legislative history, there have been constant delays in application. Equally significant, the justifications for continued use of deadlines have shifted over time. The original argument that protection of the public health was an urgent matter was undercut by the problems of identifying thresholds and incurring high costs. Similarly, the suggestion that deadlines were necessary to adopt control technologies was undercut by weak enforcement. As a result, it seems fair to say that "the deadline has become a hollow symbol in search of a tenable underpinning" (Melnick 1984, 128).

Ironically, the conventional wisdom as well as a variety of studies agree that overall air quality has improved, at least in part as a consequence of these programs (Conservation Foundation 1984; Editorial Research Reports 1982). But accomplishments have not come without costs. In fact, the problems associated with action-forcing and technology-forcing approaches can be traced largely to implementation and compliance costs. According to Downing (1983, 584):

Implementation and enforcement of pollution control regulations is based upon cooperative negotiation between the source and the agency. The agency is inhibited from being tough on emitters by its political vulnerability and the cost of enforcement actions through the courts. At the same time, pro-control political forces constrain the agency from being lenient with emitters. Faced with these constraints, the agency finds it advantageous to bargain with the sources. Technical uncertainties provide a reason for such bargaining. Case-by-case decision making is the result. The source also finds it useful to bargain. By maintaining a cooperative attitude the source is able to obtain delays in compliance, technological concessions, and a virtual guarantee of no penalty for noncompliance. Being listed as in compliance has political and economic benefits for the source as well.

Thus, strict enforcement of deadlines is obstructed by concerns about generating ever larger marginal economic costs to both the agency and polluters in return for smaller incremental reductions in society's pollution burden. And while improved monitoring capability, technology-based standards, and emission offsets serve to limit negotiation somewhat, on balance they make the issues over which bargains are struck more technical in character.

Criteria for Assessing Air Pollution Policy

The highly symbolic nature of air quality policy, the role of bargaining, and the primacy of politics is now widely recognized. But one of the results of such recognition is a hardening of the bargaining position of each major party-at-interest. An acceptable compromise capable of accommodating those interests and providing a basis for policy consensus has become very difficult to achieve. The current debate has three main perspectives that are organized around the most appropriate criteria by which air quality performance ought to be assessed and according to which reforms ought to be structured. These are (1) economic efficiency, (2) environmental effectiveness, and (3) social equity. To be sure, the three perspectives overlap in the positions of most interest groups and are combined in various ways in the ongoing public debate. But a strong case can be made that most actors have clear priorities among the three criteria and that in their assessments and proposals most groups have used one of them for their foundation.

Air Quality and Economic Efficiency

Advocates of economic efficiency as a criterion for assessing clean air policy argue that the costs of the CAA and its amendments far outweigh any benefits that have accrued to American society. As a result, frequently relying upon benefit-cost analyses, proponents of this perspective tend to assert that a more optimal allocation of resources for pollution control, measured either in terms of the marginal costs for controls or incremental reductions achieved, could be obtained using market instead of command-and-control approaches. For example, some representatives of industry make a case that regulatory inefficiencies are so gross that it is not possible to "fine-tune" existing programs. Before his return to EPA, William Ruckelshaus, then still with the Weyerhauser Corporation, put forth the typical industrial critique of the air quality system. In a letter to Vice President George Bush, he wrote that it was "inherently impossible" to attempt to cope with air pollution by basing national policy on health or environmental effects alone, and called the CAA "a complicated, pervasive law that is causing our society to spend very large sums of money for marginal benefits and thus should be changed" (Imperato and Mitchell 1985, 219).

Evidence cited in support of this position goes back as far as the controversial work of Murray Weidenbaum, former chairman of the Council of Economic Advisers, who estimated a total cost of government regulation in 1979 in excess of $100 billion (1977). This and other more recent studies have been highly critical of environmental health and safety regulation (see Gatti 1981). They attribute a high percentage of the total costs of regulation to EPA and other social regulatory agencies. Such studies argue that when benefits have been generated from programs such as air quality controls they "had as much to do with cycles in industrial activity as with regulation" (MacAvoy 1979, 101). Moreover, these analyses have gone beyond simple cost of compliance to the regulated industry calculations to argue that clean air regulations impose serious restrictions on U.S. productivity and innovation. As a result, environmental protection and other forms of social regulation such as occupational health and safety rules are identified as major factors in creating inflation and balance-of-trade problems (Imperato and Mitchell 1985, 218–21).

Yet, in spite of their importance, there are suprisingly few comprehensive estimates of either past or future expenditures for pollution control.[3] Sometimes the available information is conflicting or disparate. Industry estimates of compliance costs for the CAA are often controversial, and estimates from other, sometimes equally biased, sources such as environmentalists or EPA frequently show much lower costs (see Rubin and Torrens 1983). For example, both the Bureau of Economic Analysis and the Bureau of the Census in the U.S. Department of Commerce, as well as the Council on Environmental Quality (CEQ), have developed annual estimates of expenditures on pollution control. Similarly, McGraw-Hill, Inc., has conducted annual surveys of firms to establish estimates of actual or planned capital expenditures for pollution abatement. Typically, there are considerable differences among estimates even for actual or historical expenditures, and the variation is even greater for estimates of planned expenditures. While within the industrial sector differences exist, the McGraw-Hill estimates are higher than those of the government agencies. Such differences stem primarily from the fact that each estimate is based upon different sample sizes, composition, and assumptions about the type and cost of equipment used to comply with environmental regulations. In addition, although empirical evidence is lacking to support such an assertion, individual firms may deliberately report erroneously high costs in order to cast environmental regulation in a negative light (see Sonstelie 1981).

Industry interests, nevertheless, have used calculations of the costs of pollution control such as those in table 4 to press for extensive modifications in the Clean Air Act. Because clean air regulations affect individual firms and industrial sectors differently, a specific company or even an entire industrial sector may have a relatively limited reform agenda. But when aggregated, private-sector complaints cover almost every aspect of the air-quality control system. Particularly troublesome, from the industry viewpoint, are the NSPS and PSD standards, which are seen as major barriers to economic activity.

A second, more moderate objection to the existing regulatory structure under the CAA is advanced primarily by environmental economists. Based largely in academia or nonprofit organizations, they have attempted to link efficiency concerns with effectiveness issues (see Magat 1982). The second criticism based on economic efficiency acknowledges regulatory-

TABLE 4
National Expenditures for Pollution
Abatement and Control, 1972–1983

	Expenditure Category		Air as % of Total
	Total	Air Pollution	
1972	18.434	6.482	35.1
1973	20.603	7.832	38.0
1974	21.307	8.092	37.9
1975	23.008	9.119	39.6
1976	24.325	9.546	39.2
1977	24.800	9.805	39.5
1978	26.330	10.185	38.6
1979	26.936	10.749	39.9
1980	26.353	10.917	41.4
1981	25.536	11.463	44.8
1982	24.304	10.836	44.5
1983	25.182	11.499	45.6
$\bar{X}$	23.926	9.710	40.3

Source: Council on Environmental Quality (1985).
Note: Expenditures are given in 1972 constant dollars in billions by business and government.

generated improvements in air quality, especially advances in auto emissions control and new plant performance, and it is highly skeptical of assertions that environmental regulations have caused major problems in industrial productivity or innovation. But relying heavily on economic theory and analysis, a number of observers have rejected the conventional wisdom that enormous progress in air quality can be attributed to environmental regulation since the CAA was enacted (see Crandall 1983a; Lave and Omenn 1981). Rather, they credit fuel switching and the stagnant economy of the 1970s for most of the improvements in air quality and downplay environmental benefits associated with government intervention. As a result, they suggest a host of market or quasi-market alternatives to command-and-control regulatory systems. These include marketable pollution rights or discharge licences, emission fees, and expansion of EPA's use of the "bubble" concept for existing sources[4] and emission offsets for new sources initiatives.[5] The regulatory reform agenda also includes giving more attention to rational choice analytical techniques, such as expanded risk-benefit utilization, in the decision process.

Air Quality and Environmental Effectiveness

A second school of thought goes beyond asking questions about the relative costs and benefits of air pollution control to address goal-attainment issues. As noted earlier, there are strong links between the efficiency and effectiveness arguments made by environmental economists. Most of the proposed economic incentives, such as emission charges, are assumed to have a secondary effectiveness as well as reducing costs. But many observers raise serious questions about whether these options can achieve efficiency while at the same time cleaning the air. Again, quoting Downing (1983, 584):

> In contemplating the enforcement of transferable discharge permits or practical effluent charges, it has been argued that technical uncertainties will still exist and probably will become more crucial to the enforcement process. Likewise, we cannot expect the political setting within which the enforcement agency operates to change significantly. This implies that bargaining and case-by-case decision making will continue under these alternatives. The result will be a move away from the efficient allocations envisioned in economic writings on these alternatives.
>
> The problem of environmental externality control generates costs to some and benefits to others. Government action is required to accomplish control; therefore, environmental quality improvement is a political issue. Changing institutional forms will not make it less political. We cannot take the enforcement agency out of a political situation which requires negotiation. We cannot establish a system which penalizes emission violations at rates sufficiently high to ensure compliance. If we could, the current regulatory system could be made more effective. The politically and legally active pro-control lobby has not been able to do this for the existing system, and it is not likely to be more than marginally successful if a new institutional form is adopted.

In other words, implementation problems affect market-incentive approaches as well as those approaches based on more legal-administrative frameworks.

Equally important, from the perspective of those concerned with effectiveness, is the fact that a market-oriented approach "does not necessarily involve an orientation toward goal achievement" (Marcus 1980, 298). While a market-oriented approach may lead to a reduction in pollution, the location and amount of reductions achieved, if any, depends upon the economic efficiency of such actions. In fact, under some conditions, it may be desirable, from an economic standpoint, for an individual firm to continue to pollute the air, and reductions may not be achieved where they are most effective from an environmental standpoint.

Which organizations or interest groups represent the demand for environmental effectiveness? A good example is provided by the National Commission on Air Quality (NCAQ), established by Congress to review the CAA before reauthorization. This body concluded that the benefits of air quality control were between $4.6 and $51.2 billion annually, at a cost of about $16.6 billion per year. Based on this estimate, the NCAQ recommended the continued use of all standards that had been established, without consideration of costs. But it also proposed some 109 changes in the legislation, most of which had to do with improving effectiveness. These included recommendations to increase funding for enforcement at all levels of government, more money for R&D and for developing means of control, higher penalties for noncompliance, and strengthening transboundary pollution rules. Efficiency certainly was not forgotten in these recommendations—continuing the SIP requirement to consider costs was a key element of the report. However, on balance, the NCAQ appears to have tried to focus attention on making the national air quality system meet its goals, rather than emphasizing net benefits (National Commission on Air Quality 1981).

A similar position is found in the President's Commission for a National Agenda for the Eighties (1980, 41), which objected to the efficiency advocates' use of cost-benefit criteria in air quality management:

Whatever the monetary costs and benefits may be, environmental policy issues cannot be reduced to economic equations. Claims and counterclaims about economic impacts couch the issues in narrow and contrived terms, as if the critical question were whether an environmental standard can be justified by its potential return on investment. In fact, there is no objective way to measure these standards' total benefit to society and to the environment.

Effectiveness advocates, typically representing governmental, scientific, and technical elites, urge fine-tuning and adjusting the air quality management system as a primary emphasis for policy options. Implicitly, these parties-at-interest seem to recognize that substantial tradeoffs are a necessity in public policy. As a consequence, they generally advocate reforms that try to balance efficiency and equity concerns in an incremental fashion.

Air Quality and Social Equity

Interests concerned with improving the environment as a matter of social equity typically emphasize symbolically important values such as fairness, justice, and openness in their approach to assessing air quality control efforts. Most environmental groups that participate actively in the ongoing policy debate reflect this perspective. They often assert with an almost religious fervor that there should be no right to pollute even if the polluter pays. As such, these participants in the policymaking process tend to articulate a broad, social equity orientation, rather than a more narrow focus on economic equity.

Advocates of social equity make a three-pronged assault on critics of the existing regulatory framework. First, they charge that benefit-cost analysis, the tool emphasized by economic efficiency interests, systematically downplays distributive justice, improperly discounts the future, and undervalues environmental factors. They argue that air pollution generates substantial costs and that these burdens are regressive. Second, they typically assert that economic efficiency often is a euphemism for deregulation, since in almost any environmental policy debate, costs will be more readily identifiable and more concentrated, while benefits will be less well known and more diffused. Third, these individuals and groups are highly critical of the economic assumption that the purpose of regulation is efficiency in allocating resources. Instead, they argue that "equity is far more likely to be the *actual* goal of regulation than efficiency; that is, an analyst is more likely to be able to explain why regulation exists in an area by referring to equity than by referring to efficiency" (Meier 1985, 257).

Social equity advocates generally assert that the existing

air quality system works relatively well. Typical of this line of defense is the following:

> It is an enormous task to oversee the 160 million tons of air pollution emitted annually, the 225 million tons of hazardous waste generated, the 4 million tons of toxic chemicals discharged into waterways and streams, and the 55,000 existing chemicals, plus the 600 to 1,000 new ones added each year. Aided over the years by the development of a strong bipartisan consensus in Congress and the public's concern about environmental issues, the EPA has made some progress toward environmental cleanup during its short lifetime. Studies show that our air is significantly healthier than it was ten years ago, while numerous rivers that were pronounced "dead" twenty years ago are returning to life. Water quality degradation has slowed down, despite increases in population and industrial production. With these success stories, and growing recognition of the relationship between environment and health, public support has remained strong. Ninety-one percent of the public supports a strong Clean Water Act, and 78 percent supports a strong Clean Air Act. (Claybrook 1984, 118).

Naturally enough, most of the advocates of a social equity position represent environmental and consumer groups. But, as the above passage indicates, these groups can claim strong backing from the general public. Yet while broad support for clean air appears to be a constant, it is difficult to mobilize that support and bring it to bear on actual policy choices rather than general issues. Thus, like their counterparts in industry, environmental interests often appear to distrust the bargaining process in air quality regulation. However, unlike economic efficiency advocates, who oppose negotiation because it may be sloppy and somewhat unpredictable, social equity advocates maintain that they fear the disproportionate political power in the hands of industry and its allies. This has been the case especially with regard to bargaining arrangements under Reagan. Unlike its predecessors, the Reagan administration is seen to have mounted a concentrated and deliberate effort to dismantle regulatory programs in the interests of industry profits (Lash, Gillman, and Sheridan 1984).

Thus equity interests generally favor legal-administrative approaches to achieving air quality, so-called command-and-control initiatives, many of which characterize the existing system. These arrangements, it is felt, give environmentalists visible, highly symbolic advantages in the policy process, as well as important adjudicatory fallback positions. Therefore, even deadlines, standards, and troublesome enforcement mechanisms, with all their problems, are favored by most environmental and consumer interests. And these groups support the use of technological fixes, like the scrubber, because they seem to promise more uniform and certain compliance. In addition, their support for the command-and-control approach often stems from their fervent opposition to any market-oriented approaches to pollution control such as emission fees or marketable rights, which might condone the right to pollute if the polluter pays. This dogmatism has been a significant obstacle to finding alternatives to command-and-control approaches.

Environmentalists do have a reform agenda, however. High on their list of proposed changes is a greater focus on the distribution of environmental costs and benefits, coupled with more attention to the global dimensions of environmental problems. Concern for the impact of acid rain is the highest priority item under this second imperative. And because social equity advocates have been the most visible actors in their efforts to get acid deposition on the national environmental policy agenda, they are widely assumed to be the only participants with a vested interest in pursuing the issue. This is not the case. As we demonstrate in the next section, each set of parties-at-interest may have something to gain from having acid precipitation dominate the clean air debate.

Acid Rain Politics: Surrogate for Clean Air or Emerging Policy Arena?

Why has acid rain received so much attention in the last few years? To be sure, existing scientific knowledge about acid deposition has advanced, albeit by something less than leaps and bounds. And our understanding of the relative contributions of various control technologies has improved. But if science and/or technology are not the only driving forces, why the focus on acid rain? One possible, and hardly radical, explanation is that the larger air quality debate had bogged down precisely as a consequence of the factors outlined above, and acid rain provides a convenient outlet for all participants.

First, consider the situation in which industry and other efficiency-oriented interests found themselves by 1982. The Reagan regulatory relief program had stalled in the face of concerted and growing environmental opposition. The Clean Air Act was not going to be rolled back after all. Ann Burford, EPA administrator, and James Watt, secretary of the interior, had, after a brief period, acquired at the national level the label they had earned in Colorado—"the crazies." As a result, the quite reasonable idea that environmental regulations developed in the optimism of the 1960s and 1970s ought to be redefined for the 1980s and beyond in the light of more than a decade's experience was destroyed by a heavy-handed, myopic approach to environmental decision making and regulatory relief.

Missed opportunities are not new in American environmental policy, but the case of air quality and acid rain may represent the height of frustration for reformers. We have noted the long history of air pollution problems and control efforts in England and our own country. Warning signs were everywhere, and the need for mitigation of local ambient problems was obvious. Yet not much was done until crises occurred. Thus, to some extent the Reagan administration's "go slow" approach on environmental matters had the appearance of the traditional U.S. government response. But this was deceiving. Under the guise of economic efficiency and "good science,"[6] the administration missed a chance to develop a new reform coalition and instead attempted to roll back most environmental regulations. Moreover, if the CAA, the cornerstone of the nation's environmental legislation, could be attacked, acid rain seemed like such a marginal issue that it was viewed almost with contempt by administration spokespersons. Typical of this attitude was the statement by Interior Secretary Watt: "Every year there's a money-making scare. This year it's acid rain" (Tobin 1984, 243).

So why should acid rain's ascendancy to the upper reaches of the national environmental policy agenda please industrial advocates of efficiency? One possible reason might be that, compared to the broader air quality debate, acid rain gives these interests a more favorable benefit-cost calculation. That is, as we noted earlier, the acid deposition issue is characterized by demonstrable costs and highly uncertain benefits, while the total air pollution control effort has substantially better evidence for its net benefits. Moreover, after a decade of environmentalism, it became more publicly acceptable for polluters to want to know details of costs and benefits before committing themselves to additional control programs. This key political factor has not been overlooked by efficiency interests. Nor have they ignored the fact that acid rain mitigation, as we shall see, may place additional economic burdens on economically depressed regions of the country and on financially troubled industrial sectors within those regions. Add to this the international dimension of acid precipitation—a substantial portion of the benefits of mitigation might go to Canada—and you have a strong macroeconomic argument in favor of caution and against expensive control schemes. As an editorial in the *Indianapolis Star* (1984, 10) put it:

With new information turning up all the time, legislation that punishes one section of the country or one segment of the economy—as the auto emissions theory does—would not only be unjust but impractical. Half solutions are as bad as hasty, biased judgments.

Whatever acid rain remedies are finally prescribed will be exceedingly expensive and Congress has a responsibility to make sure the remedies work. Acid rain theories are numerous. What is missing is irrefutable evidence regarding either sources or relative effects. Until Congress has that, it must move slowly and cautiously.

To the degree that cautious and deliberate development of policy alternatives leads to an emphasis on R&D and technological innovation, the efficiency argument finds strong support among other parties-at-interest.

But environmental groups in particular view much of this line of thought as a delaying tactic. They maintain that it is designed to postpone action until either the political balance of power shifts in their favor or the problem itself goes away through, for example, the phasing out of older, less efficient facilities. Environmental groups assert that industry and the Reagan administration have used the challenge of scientific rationality to undermine both new regulatory proposals and existing regulatory systems. By pressing for scientific proof and certainty of health effects or other adverse consequences before introducing new controls, and by seeking irrefutable evidence of clear relationships between cause and effect before taking action, economic efficiency advocates often are seen by their opponents as using science as a legitimization for deregulation (Dickson 1984). When this concern is combined with the allegations of abuse of scientific endeavors in the Reagan administration—ranging from pressures on agency scientists to alter analyses, to falsified data by researchers, to tensions between Washington bureaucrats and regional officials with regard to assessments—it naturally undercuts the administration's calls for "good science" (Claybrook 1984; Lash et al. 1984; Rushefsky 1984).

Nevertheless, a policy emphasizing R&D and technological demonstration is attractive, especially to effectiveness-oriented participants. Clearly, the scientific and technical communities have every reason to support an expanded acid rain research agenda. This is especially true given the cutbacks or stagnation in other environmental science and engineering research fields. To cite only the most obvious example, EPA's research budget for fiscal year 1985 totaled $306 million in current dollars, still far below the FY 1981 total of $385 million. And the acid rain component of the budget represented by far the most dramatic increase in this funding. The commitment of some $38 million to acid deposition R&D in FY 1985 more than doubled the level of federal support for EPA of the previous year. Compare this to the funding pattern for the air program, which received an increase of 4.7 percent over its FY 1984 level, to a total of $66 million (American Association for the Advancement of Science 1986, 93–100). The numbers for FY 1986 are even more striking. EPA's research and development program for that year was $325 million. That represents a 6 percent increase from the previous year but is still some $60 million shy of the FY 1981 level. Additional emphasis on the causes and effects of acid rain once again was the most significant winner, with a budget expanded to $47 million. A similar pattern emerges when one considers the steady increase in acid rain R&D funding across federal agencies and departments.

Moreover, the priorities of the acid rain program have become broader over time. Beginning with an emphasis on aquatics research and defining existing scientific knowledge, the program has expanded to include major field studies and complex computer modeling of pollutant transport and chemical transformation, quantification of air concentration levels for acid deposition precursors, a nationwide precipitation monitoring system, baseline surveys of lake chemistry, surveys of forest damage, analyses of acid fog, dry deposition monitoring, upgrading emissions inventories, developing models to predict costs, and testing control technologies (Hollander 1983). More recently, increased emphasis on field testing of materials damage has been added to the program. This is a huge agenda. When combined with the rapid increases in R&D funding levels, it is no surprise that individuals associated with the scientific and engineering components of the environmental effectiveness-oriented community, even those who personally may support emissions reductions, have been enthusiastic supporters of the Reagan administration's emphasis on acid rain R&D.

Environmental groups and other advocates of the social equity perspective, on the other hand, point to inconsistencies between the Reagan administration's attempts to weaken the CAA and its support for acid rain R&D. They also are highly critical of what they see as the attempt to use administrative rule-making processes to undermine clean-air programs (see Friends of the Earth 1982). But for these parties-at-interest in the policy debate, acid rain also holds substantial promise. Many environmental interest-group victories in the past have come as a result of dramatic environmental crises or, at least, the public awareness of such crises. From the concern with DDT in the food chain to the hazardous waste problems at Love Canal, equity interests have been mobilized by what has been termed entrepreneurial politics. This brand of decision making is a function of the emergence of a skilled entrepreneur who can transform largely latent public attitudes into action by capitalizing on a scandal or emergency (Kingdon 1984; Polsby 1984; Wilson 1980).

Acid rain certainly has the potential for such an entrepreneurial situation. In fact, many argue that an emergency already exists. They believe there is evidence linking acid rain to the deterioration of spruce trees on Camel's Hump Mountain in Vermont, for example. But, thus far, the perception of an immediate ecological crisis has not emerged in the broader body politic or the scientific community as a whole. Yet acid rain is the only air quality issue that has demonstrated the Reagan administration's vulnerability to date, as reflected in the R&D budget and policy rhetoric. As such, it will doubtless continue to fascinate environmental interest groups operating from an equity orientation. However, these groups have not made a strong enough argument to transform the acid rain debate dramatically enough to make consumers in California or North Carolina willingly subsidize Midwest polluters causing acid rain in New England or Canada.

Debates over the relative costs and benefits of acid rain mitigation, support for R&D initiatives, and lack of consensus about the urgency of government action are exactly what one would expect. Considering that acid rain serves as a surrogate for the broader air quality conflict, the absence of such debates would be astonishing. Efficiency, effectiveness, and equity interests alike appear to have good reason to try to inject acid deposition issues into this conflict, as each set of actors tries to break the logjam that has characterized air pollution policymaking for the past eight years. But as long as acid rain is treated as just another dimension of the air quality debate, it seems unlikely that new political alliances will be forthcoming. The 1977 "clean coal/dirty air" coalition, made up of defenders of declining "frost belt" industries, high-sulfur coal producers, and environmentalists, was from the first an "unholy alliance" that was hardly an irresistible political force. In Ackerman and Hassler's terms (1981, 117), it was a "fragile coalition depending for its survival on peculiar conditions." We have outlined these conditions in previous sections. Such a vulnerable combination of forces could not be counted upon to serve as a solid foundation for policy formulation over any extended period, and in fact it has not. To the extent it has affected environmental decision making, the alliance has defended the status quo.

Unfortunately, acid rain as a surrogate for the larger air quality issue does not appear to offer any politically feasible alternatives for coalition formation. But there is another side to the acid rain question—factors that are unique to the acid deposition problem and that may affect environmental politics in quite unpredictable ways. Acid rain may, in fact, be evolving into a policy issue that causes us to reappraise our clean air concerns and underlying approach to ambient air quality management.

The 1977 CAA amendments clearly encouraged political forces in the older industrial areas of the country to form a coalition with environmentalists. However, the local coal use and forced scrubbing provisions of the 1977 amendments worked against such a coalition. In fact, a desire to maintain existing coal market shares and avoid further reductions, especially by retrofit requirements for existing plants, provides a strong incentive for midwestern interests to oppose the proposals of their former coalition partners for addressing international, interstate, and interregional air pollution. Thus, acid rain as an independent policy concern is a double-edged sword. On the one hand, current proposals to address the long-range transport of sulfur oxides make it highly improbable that the old alliance can survive; indeed, odds are that some new framework may emerge. On the other hand, because of the politics of acid rain, it is difficult to see how any other coalition can be developed to overcome the current inertia in the system.

2

THE SCIENCE OF ACID RAIN: AN EVOLVING CONSENSUS?

Perhaps, when all is said and done, it is not really so remarkable that acidification could go unnoticed for years—right up to the end of the 1960s. In contrast to environmental influences of many other kinds, acidification is a furtive process—in its early days almost unnoticeable. Our senses of smell and taste are not capable of distinguishing between acidified and unaffected lake or well water. The clear limpid water in an acidic forest lake can also, in many cases, lend it a deceptive beauty. And the trees growing in an acidified forest area look just like trees anywhere else, at least as long as the acidification is moderate.

—*Swedish National Environmental Protection Board (1981)*

WHAT IS ACID rain and why should we be concerned about it? Acid rain commonly refers to what is identified more precisely as the wet and dry processes for the deposition from the atmosphere of acidic inputs into ecosystems. Thus, all forms of precipitation—not just rain—can be acidic. Indeed, the definition includes acidifying compounds that are deposited in dry form. As a result, *acid deposition* is the scientifically accurate and all-encompassing term for acid rain. For simplicity as well as by conventional usage, however, the term *acid rain* is commonly used to include both precipitation and dry deposition.

Since acids release hydrogen ions in a water solution, the relative acidity or alkalinity of any solution is typically described by the percentage of hydrogen ions that a water solution contains measured on the logarithmic potential hydrogen (pH) scale.[1] Hydrogen ions have a positive electrical charge

FIGURE 1
The pH Scale

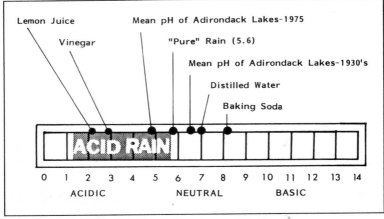

Source: U.S. Environmental Protection Agency (1980).

and are called *cations*. Ions with a negative electrical charge are known as *anions*. A substance containing equal concentrations of cations and anions so that the electrical charge is balanced is *neutral*. As shown in Figure 1, a solution that is neutral—that is, neither alkaline (base) nor acidic—has a pH value of 7.0. A substance with more hydrogen ions than anions is acidic and has a value of less than 7.0 on the pH scale. Substances with more anions than cations are alkaline and have a pH that measures above 7.0 on the scale. Thus, as the concentration of hydrogen ions increases, the pH decreases to represent greater acidity. The further a reading is from 7.0, below or above, the more acid or base the substance is. Because pH expresses the negative logarithm of acid concentration, interpreting changes in chemical composition can be confusing for many laypersons. The lower the pH value, the higher the acidity. Each full pH unit drop represents a tenfold increase in acidity. For example, a solution whose pH value equals 4.0 contains ten times as much acid, not just 20 percent more, than one measuring 5.0 on the pH scale. And, it is one hundred times more acidic than a substance with a pH of 6.0.

All forms of precipitation—rain, snow, sleet, hail, fog, and mist—are naturally somewhat acidic, and human activities ve made them more so. For example, in industrial regions, the pH of rainfall is often around or below 4.0, and it has been measured as low as 2.6. In pure water, the "natural" acidity value often is assumed to be pH 5.6, calculated for distilled water in equilibrium with atmospheric carbon dioxide concentrations. This is a somewhat arbitrary value. The presence of other naturally occurring substances such as sulfur dioxide, ammonia, organic compounds, and windblown dust can produce "natural" values ranging from pH 4.9 to 6.5 (Charlson and Rodhe 1982; Galloway et al. 1982). Table 5, however, reveals that natural sources such as lightning, microbial activity in soils, and biogenic processes make relatively small contributions to total nitrogen and sulfur emissions in the United States. The ratio of anthropogenic to natural sulfur emissions is on the order of at least 3:1 and it is over 7:1 for nitrogen oxide emissions.[2] Thus, because manmade rather than natural sources for the sulfur and nitrogen oxides released into the atmosphere for conversion to acids predominate in North America and Europe, the atmospheric chemistry suggests that reducing anthropogenic emissions of acid rain's major causes should reduce the aggregate level of acidic deposition, although it may have less impact on some sensitive receptor areas than popular impressions imply.

TABLE 5
Relative Contribution of Natural and Anthropogenic Sources to Total Sulfur and Nitrogen Oxides Emissions in the United States

	Estimated Tg's per year[a]
Sulfur Emissions	
Biogenic	0.10 (upper bound)
Marine	0.36 (upper bound)
Volcanic	0.20 (upper bound)
Anthropogenic	1.70 (1981 level)
Nitrogen Oxides Emissions	
Fossil fuel combustion	5.80
Biomass burning	n.a.
Lightning	0.30
Microbial activity in soils	0.40
Oxidation of ammonia	n.a.
Stratospheric inputs	negligible
Total for natural sources	0.80

Source: Roth et al. (1985).
a. 1 Tg equals one million metric tons.

Robert Angus Smith, a nineteenth-century English chemist, might well lay claim to the title of being the "father of acid rain." In his pioneering studies of precipitation chemistry and its effects, Smith first used the term *acid rain*. Drawing upon data measuring the chemistry of rainfall from England, Scotland, and Germany, Smith demonstrated that variation in regional factors such as coal combustion, wind trajectories, the amount and frequency of precipitation, proximity to seacoasts, and the decomposition of organic materials affected sulfate concentrations in precipitation (Smith 1872). His efforts of more than a century ago startlingly mirror the contemporary research agenda. For example, he established a network for collecting and analyzing precipitation samples, discovered that the acidity of rain was dominated primarily by its sulfate content, and, perhaps most interestingly, speculated about whether damages to trees and crops were attributable to acid rain, to the direct actions of gaseous pollutants, or to natural factors. Smith's work, however, was largely ignored and failed to generate immediate follow-up research in spite of the fact that he was the inspector-general of the Alkali Inspectorate for the United Kingdom.[3]

Cowling (1982) asserts that contemporary concern about acid deposition and its effects originated in three seemingly unrelated areas: limnology, agricultural science, and atmospheric chemistry. Svante Oden (1968), in the first major attempt to integrate knowledge from those disciplines, maintained that analyses of air mass trajectories matched to temporal and spatial changes in precipitation chemistry indicated that sulfur and nitrogen were transported long distances, ranging from 100 to 2,000 kilometers. Oden also asserted that clearly identifiable source and receptor areas existed. His analysis formed the initial basis for concluding that acid deposition is a large-scale regional phenomenon with long-term adverse ecological consequences. Thus, as figure 2 indicates, information produced by a complex and rapidly evolving body of research forms the scientific basis for defining the acid deposition problem.

An Overview of Existing Knowledge

Acid rain first emerged as a public policy concern at the 1972 United Nations Conference on the Human Environment in

FIGURE 2
Schematic View of the Acid Deposition Problem

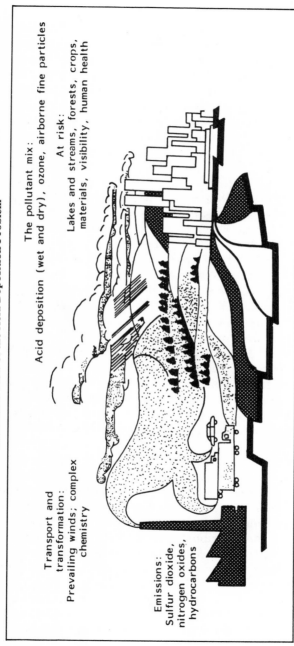

Source: U.S. Office of Technology Assessment (1984).

Stockholm. In a case study prepared for the Stockholm conference, Swedish scientists asserted that precipitation acidity attributable to SO_2 emissions from manmade sources, primarily industrial processes and utilities, was causing adverse ecological and human health effects (Swedish Ministry for Foreign Affairs, Swedish Ministry of Agriculture 1972). Largely in response to the Swedish study, a number of major research efforts were initiated in other countries to address the causes of acid deposition, as well as its direct and indirect ecological effects.

The first of those endeavors to identify adverse ecological effects began in Norway in 1972. The Norwegian Interdisciplinary Research Program, commonly referred to as the SNSF Project, focused on establishing the effects of acid precipitation on forests and fish (see Overrein, Seip, and Tollan 1980; Drablos and Tollan 1980). The SNSF Project, entitled "Acid Precipitation—Effects on Forest and Fish," was the largest multidisciplinary study in Norwegian history. Its annual budget was approximately 10 million Norwegian kroner, equivalent to U.S. $2 million, from 1972 to 1980. The SNSF Project involved cooperative research undertaken by 12 Norwegian institutions and more than 150 scientists. It produced two major international scientific conferences. The first was held midway through the study at Telemark, Norway, in June 1976 (see *Ambio* 1976; Braekke 1976). The second conference at Sandefjord, Norway, in March 1980 concluded the SNSF Project and provided a forum to evaluate the state of existent knowledge (see Drablos and Tollan 1980; Overrein, Seip, and Tollan 1980).

In 1972, another major research project, this one in the atmospheric sciences, was undertaken to address the problem of linking source and receptor areas. Acting under the auspices of the Organization for Economic Cooperation and Development (OECD), eleven European nations launched a cooperative effort to measure the contribution of local and transboundary sources to each participating country's sulfur deposition (OECD 1977). Austria, Belgium, Denmark, Finland, France, the Federal Republic of Germany, the Netherlands, Norway, Sweden, Switzerland, and the United Kingdom actively participated in the study. Italy participated on a more limited basis in some of the data collection. Data collected by aircraft sampling and at seventy-six ground monitoring sites for the European Air Chemistry Network (EACN) were reported monthly to the Norwegian Institute for Air Research, which coordinated the study (Ottar 1976). Transfer coefficients derived from long-range transport models that simulate atmospheric processes were used to estimate the relationship between source emissions and receptor deposition levels (see Regens and Donnan 1986). The OCED study concluded that SO_2 emissions could be transported long as well as short distances. In five of the eleven countries participating in the study, more than 50 percent of total sulfur deposition was estimated to come from nondomestic sources (OECD 1977).[4] However, because of serious problems with the reliability of national emissions data as well as the accuracy of atmospheric transport models for estimating site-specific deposition inputs, the OECD study's findings are subject to plus or minus 50 percent error for individual receptor estimates. Nonetheless, while meteorological changes and more accurate emissions inventories could alter significantly an individual country's contribution, the study reinforced the conclusion that some proportion of the acid deposition occurring over almost all of northwestern Europe is due to transboundary pollution.

During the early 1970s, studies conducted in Canada and the United States produced similar concerns about the possible environmental consequences of acid deposition (see Cogbill and Likens 1974; Beamish and Harvey 1972; Likens, Bormann, and Johnson 1972). Thus, by the mid-1970s, studies noting declining pH and speculating about the possible impact of acidification on aquatic and terrestrial ecosystems had been reported in Sweden, Norway, Canada, and the United States (Cowling 1981, 1982). Since the mid-1970s, other Western European countries such as the United Kingdom, the Federal Republic of Germany, the Netherlands, and Austria have become increasingly concerned as the potential impacts of acid deposition have been identified in both their own and neighboring states (see Ulrich 1982; Wright et al. 1980; Vermeulen 1978).

Chemistry of Acid-forming Compounds

The major contributions to acid rain are sulfur oxides, primarily sulfur dioxide and to a lesser extent sulfur trioxide

(SO_3), and various oxides of nitrogen, principally nitrogen dioxide (NO_2) and nitric oxide (NO), commonly labeled NO_x. When a fossil fuel is burned, the sulfur and nitrogen in the fuel combine with oxygen in the atmosphere to form sulfur and nitrogen oxides. In contact with air, SO_2 and NO_x are spontaneously oxidized to form sulfate and nitrate:

$$SO_2 + \tfrac{1}{2} O_2 + H_2O \rightarrow 2H^+ + SO_4^=$$
$$NO + NO_2 + O_2 + H_2O \rightarrow 2H^+ + 2NO_3^-$$

While the overall chemistry leading to the formation of acid rain is reasonably well defined, important questions remain regarding the underlying dynamics of that process (see Jacobs 1986; Hoffman 1986). For example, defining the reaction pathways—sometimes termed transformation processes—for SO_2 and NO_x and establishing the conditions under which specific pathways produce acidic compounds is important in evaluating the efficacy of various emissions reduction scenarios. If the amount of sulfate produced by a pathway is proportional to the amount of sulfur dioxide in the atmosphere, then reducing the level of sulfur dioxide emissions will reduce proportionately the amount of sulfate formed. But, if the reaction process is oxident-limited, then the availability of ozone may become the dominant pathway for acid formation. Overall, however, the conversion of sulfur dioxide to sulfate is completed within a period ranging from several hours to several days, while the transformation of nitrogen oxides into nitrate probably occurs much faster and is completed within hours (see National Research Council 1983).

Emissions, Transport, and Deposition

Unfortunately, the simplicity of the term *acid rain* also conveys the image of an easily measured and understood phenomenon. In fact, the acid rain problem stems from a series of complex and varied chemical, meteorological, and physical interactions (Regens 1984). For example, monitoring programs have focused on measuring only the precipitation (wet) component. Values for total acid deposition, which includes a dry and wet component, consist of estimates because valid and reliable methods for directly measuring the dry component on a field network basis are still being developed (see Bunsinger 1986; Giorgi 1986).

Continuous, large-scale precipitation monitoring has been conducted in Scandinavia since the 1950s. But, with the exception of the Hubbard Brook Experimental Forest in New Hampshire, continuous North American precipitation monitoring data have been available for less than a decade. As a result, long-term trends for the United States and Canada are poorly defined. Trend studies for North America have been subject to considerable controversy because of differences in collection methods, siting criteria, chemical analysis techniques, sample storage methods, and quality assurance. The lack of a sufficient number of continuous sampling sites also is a source of possible error (Regens 1984). Bearing these limitations in mind, we can examine current levels and trends (see Federal/Provincial Research and Monitoring Coordinating Committee 1986).

Precipitation chemistry data collected in the 1970s and 1980s indicate that acid deposition occurs throughout eastern North America. The area of greatest acidity is concentrated over eastern Ohio, western New York, and northern West Virginia (U.S.-Canada Work Group 2 1982). Lewis and Grant (1980) found that rural areas in the western United States at a higher elevation also have low pH rainfall, probably due to significant increases in nitrogen oxide emissions throughout the region.

Broad geographical generalizations derived from such individual monitoring site data are extremely imprecise. For example, the final report on atmospheric sciences prepared by Work Group 2 of the U.S.-Canada Memorandum of Intent on Transboundary Air Pollution (MOI) concluded that individual monitoring station values were reasonably accurate. Work Group 2 also evaluated the uncertainty in isopleth map lines generated from site data.[5] The MOI report concluded that the degree of possible error in those isopleths was about plus or minus 20 percent in magnitude and 50 to 200 kilometers in position for eastern North America (U.S.-Canada Work Group 2 1982). The degree of uncertainty would be higher for western North America because fewer monitoring stations are located there.

While obvious uncertainties exist, figure 3 suggests the potential for near continental-scale impacts if adverse effects are linked to loadings of acidic compounds above a given level or

FIGURE 3
Annual Mean Value of pH in Precipitation in the United States and Canada, 1980

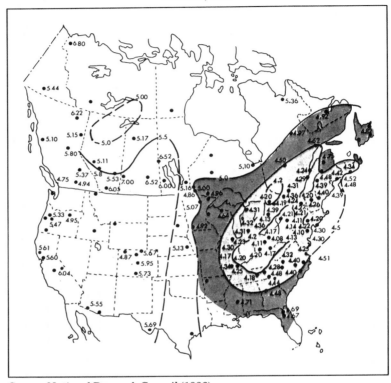

Source: National Research Council (1986).
Note: Figures are weighted by the amount of precipitation in the United States and Canada.

rate, assuming those loadings are not neutralized once they fall to earth. Initial examinations of alkalinity levels in U.S. lakes reinforce such a view. An inventory of sensitive but not necessarily altered lakes conducted by EPA reveals that more than 10 percent of the lakes surveyed in the eastern United States have pH levels of 6.0 or below, with the Adirondacks and the upper Midwest having the largest proportion of lakes with low acid neutralizing capacity (see Linthurst et al. 1986; Omernik and Griffith 1986).[6] Similarly, EPA's companion survey of lakes in the West revealed that while fewer than 1 percent of the 10,400 lakes in California, the Pacific Northwest, and the Rocky Mountains have become dangerously acidic, more than two-thirds of the region's lakes had limited acid neutralizing capacity (Marshall 1987).

One cannot assume automatically that acids are not neutralized in some potentially sensitive areas. Field studies suggest that this may occur in some alpine and subalpine lakes of the Sierra Nevada mountain range in California (see Melack et al. 1985). Even in areas whose hydrology and geochemistry fail to neutralize acid rain inputs completely, it is important to consider the interaction of acid rain with the acidifying effects of natural soil formation processes and long-term changes in land use practices when establishing the geographical extent of actual or potential damages (Krug and Frink 1983). Nonetheless, given the acidity levels of precipitation, a policymaker might want information about the possibility of controlling the sources of acid rain (see Regens and Donnan 1986).

For eastern North America, sulfur dioxide and nitrogen oxide emissions from manmade sources are estimated to be at least ten times greater than those from natural sources such as ocean-land fluxes and vegetation (U.S. Environmental Protection Agency 1983). Clearly, subject to technological and economic constraints, emissions from manmade sources such as electric utilities, industrial boilers, and motor vehicles can be limited through fuel switching, precombustion cleaning and/or postcombustion flue gas desulfurization. And because the pH of rainfall is more acidic than one might expect due to natural processes, the monitoring data encourage policymakers to seek information about the geographical distribution of manmade sources of precursor emissions. Figure 4 reveals that the areas receiving the greatest yearly averages of acid deposition are found within or downwind from, and relatively proximate to, regions containing major manmade emissions sources.

Tall stacks at a number of point sources, especially electric utilities, contribute to the likelihood of chemical conversion and the long-range transport of these emissions as acidic compounds. The taller the stack from which the pollutants are

FIGURE 4
Sulfur Dioxide and Nitrogen Oxides Emissions in the United States and Canada, 1980

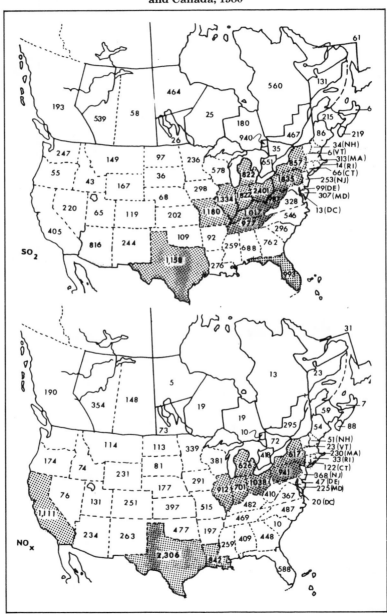

Source: U.S.-Canada Work Group 3B (1982).
Note: Figures are representative values.

emitted, the higher the gases rise into the atmosphere. This reduces ambient concentrations in the vicinity, but increases the potential for those pollutants coming to earth elsewhere. Table 6 indicates that an estimated 429 stacks taller than 200 feet were built in the United States between 1970 and 1979. Most of the tallest, those in excess of 800 feet, were built in the Midwest and Southeast (Council on Environmental Quality 1980, 175). Prevailing winds from those regions typically move in a northeasterly direction toward New England and Canada. This is consistent with the view that manmade, or *anthropogenic* sources, contribute overwhelmingly to the emissions of acidic precursors.

As a result, assumptions about the relationship between emission sources and receptor areas form the basis for any acid rain control strategy (National Research Council 1986; Venkatram 1986). And because public policies represent a response to perceived problems, this raises the question of the actual or potential adverse effects of acid rain on human health and the environment. As Huber notes (1984, 57) "Surely [the] most important question is why anyone cares whether rain, lake or soil acidity is changing." The most obvious, but not necessarily the sole answer, is that if socially valued resources are at risk and damage can be prevented, reduced, or mitigated, then action should be considered.

TABLE 6
Estimated Number of Tall Stacks
Constructed in the United States, 1970–1979

Stack Height (in feet)	Electric Utilities	Smelters	Pulp and Paper Mills	Steel Mills	Oil, Gas, and Chemical Facilities	Total
≧ 800	34	2	—	—	—	36
700–799	37	1	—	—	—	38
600–699	45	1	1	—	—	47
500–599	55	0	—	—	3	57
400–499	55	0	2	—	4	61
300–399	34	1	12	2	11	60
200–299	23	3	34	4	65	129
Total	283	8	49	6	83	429

Source: Council on Environmental Quality (1980).

Environmental Effects

Acid deposition and/or sulfur and nitrogen oxide emissions are said to affect ecosystems and human health both directly and indirectly. As figure 5 reveals, such emissions can have direct effects, such as the acidification of lakes and streams, plant damage, or reduced forest growth, as well as indirect effects on human health or reduced visibility (see Lefohn and Brocksen 1984). Critics of additional controls, however, maintain that the "huge acid rain research effort provides ample data showing that the link between SO_2 emissions and the acidity of rain is far weaker than generally supposed, and, further, that the link between acid rain and ecological damage is even weaker, or nearly nonexistent" (Katzenstein 1986, 32). What then, do we know about the environmental impact of acid rain?

Conclusive evidence points to chemical and biological changes, including fish kills, in lakes and streams that have limited capacities to neutralize acidic inputs. This can affect sport fishing, tourism, and other values associated with aquatic resources. As Crocker and Regens point out, however (1985, 114), the "current economic consequences of these effects are small relative to the economic value of all freshwater sport fishing in North America, and estimates (even with order-of-magnitude errors) of the value of current effects on other categories. Too many substitute lakes and too many alternative outdoor recreational opportunities exist." This underscores current limitations in our knowledge of the magnitude of aquatic damages and the controversy that can accompany efforts to establish quantitative estimates of the value of benefits.

Evidence of damages to nonaquatic ecosystems, especially forests, is largely circumstantial. In part, this reflects limited research. It also reflects the often synergistic nature of effects attributable to air pollution in the ambient environment, which is more analogous to a chemical soup than individual, discrete pollutants exerting effects in a noninteractive fashion. As a result, impacts are plausible, and evidence of nonaquatic effects, especially on forest productivity, is growing. For example, adverse effects on forests may result from the leaching out of soil

FIGURE 5
Potential Direct and Indirect Effects of Acid Deposition and/or Sulfur and Nitrogen Oxide Emissions

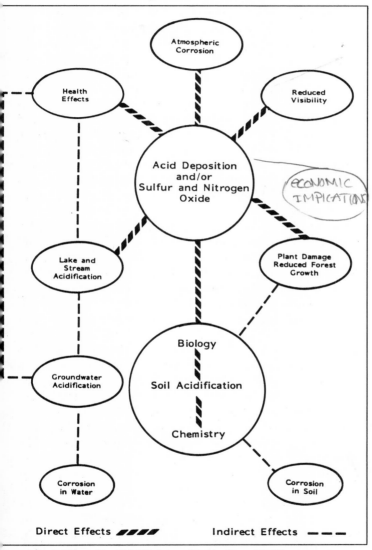

Source: Adapted from Swedish Ministry of Agriculture (1982).

nutrients[7] or through the mobilization of toxic metals (see Friedland, Johnson, and Siccama 1986; James and Riha 1986; Ulrich 1982).

As a result, many are concerned about the harmful, long-term effects of acid deposition on trees—particularly spruce, pine, aspen, and birch. Vogelmann (1982) asserts that studies of mature forests in the northeastern United States indicate reduced growth patterns as well as increased mortality for primarily coniferous species in recent decades (see also Wetstone and Foster 1983). Because causal linkages are complex, conclusions about forest effects remain somewhat equivocal. However, acid deposition does appear to be one of various stresses affecting forest ecosystems, although obviously acid rain is not the only potential culprit. A combination of atmospheric pollutants—acidic deposition, sulfur dioxide, ozone—as well as drought, temperature shifts, pathogens, and heavy metals like lead have been implicated.

Unlike the effects on aquatic ecosystems and the possible effects on forests, no clear evidence of a direct link between ambient levels of acid deposition and injury to agricultural crops has been demonstrated (U.S. Office of Technology Assessment 1984). The emerging consensus that acid deposition fails to damage crops is not surprising, since farmers use agricultural practices to manage acidic and alkaline inputs to their soils.

Concern has also been expressed about the impact of acid deposition on outdoor sculpture, historic monuments, buildings, and other structures. Environmentalists refer frequently, because of their emotional appeal, to the damage to priceless artifacts like the Parthenon or the Statue of Liberty. Damage to the Acropolis has been traced primarily to emissions from nearby traffic and from an oil refinery only a few kilometers away. The Statue of Liberty sits in New York Harbor, where the sea water typically measures around pH 8.2 to 8.4, which is highly alkaline. Its damage appears to have resulted from failed insulation that allowed electrolytic corrosion to take place between the statue's copper sheathing and iron framework, as well as from a century's exposure to corrosive sea salts (Burroughs 1984, 7). As a consequence, while field studies have linked materials damage to air pollution, such damage is most prevalent in urban areas with high concentrations of ambient sulfur dioxide. This suggests the impact of primarily local rather than distant sources.

Finally, unlike respirable sulfates or fine particulate matter, acid deposition does not appear to represent a direct threat to human health. Limited health risks may be associated with acid fog episodes or the leaching of metals such as lead into drinking water supplies, although most drinking water supply systems already have treatment facilities for liming to neutralize acid inputs, thereby preventing metals leaching.

We can use what we already know from the physical and biological sciences to construct a policy rationale for either maintaining existing emissions control programs and expanding R&D or initiating additional control measures (see table 7). First, although the extent as well as the rate at which damage occurs remains uncertain, the widespread recognition that acid deposition is an environmental problem, combined with its salience on the policy agenda, has compelled governments to respond. Second, manmade sources are the overwhelming contributors to acid deposition in eastern North America (National Research Council 1986). While it is expensive, control technology is available to reduce emissions from those sources significantly (see Organization for Economic Cooperation and Development 1983). Third, a recent National Academy of Sciences (NAS) report concluded that reductions in sulfur dioxide emissions over a broad area for several years can be reasonably assumed to produce a proportionate reduction in annual average sulfate deposition for that area if all other factors, including climatic factors, remain constant (National Research Council 1983). Finally, other parameters of air quality in eastern North America—regional visibility, particulate matter loadings, and ambient sulfur dioxide levels—are affected strongly by the causes of acid deposition (see Poirot 1986). As a result, they are likely to be improved if atmospheric loadings of precursor emissions are reduced. This underscores the link between the acid rain question and the air quality debate. Thus, while uncertainties remain about nonaquatic ecosystem effects and site-specific changes in deposition patterns and pH within sensitive receptor areas, advocates of taking further action believe it is feasible to outline a control program focusing on sulfur dioxide reductions.

TABLE 7
Opposing Interpretations of Existing Scientific Knowledge as Rationales for Acid Rain Policy

Rationale for Deferring Additional Action	*Rationale for Taking Action Now*
• Aquatic ecosystem effects are only documented damages in eastern North America. Fish population losses are limited to a small percentage of the lakes that have been studied in the U.S., primarily in the Adirondacks.	• There are thousands of potentially sensitive watersheds throughout eastern North America whose fish populations may be threatened by acidification.
• Nonaquatic ecosystem effects theoretically are plausible but only circumstantial evidence exists. Terrestrial ecosystem findings are complex and equivocal with only limited empirical data for adverse effects on forest productivity, crops, or soils. Sensitive soils may require decades for cation depletion.	• Adverse effects on forest productivity and other terrestrial ecosystems may result from acidic deposition through mechanisms such as leaching of soil nutrients or mobilization of toxic metals. Responses are likely to be subtle and, therefore, difficult to detect prior to onset of major damages.
• The effects of acidic deposition are sufficiently ambiguous to preclude calculating a target loading rate that definitely alters aquatic or terrestrial systems.	• Sulfur appears to dominate on a long-term average basis. Aquatic responses have been shown on a limited empirical basis at deposition rates of 30 kg/ha/yr with some responses observed in the 20 to 30 kg/ha/yr of wet sulfate range.
• Nitrate often dominates the acidity released during spring snowmelts in the northeast but insufficient data are available to develop target loadings for nitrate induced water quality effects.	• The areas of highest sulfate deposition lie over and immediately downwind from the region of maximum SO_2 emissions in eastern North America.
• Existing data offer little evidence that the acidity of precipitation in eastern North America has been increasing for decades.	• For a given emission magnitude, acid deposition attributable to a source will decrease as distance between source and receptor increases.
• Chemical transformation processes are not well understood so specific source-receptor relationships cannot be defined. Current uncertainties preclude specifying an optimal spatial strategy for imposing emissions reductions.	• Existing models and empirical data for zones of influence suggest that in the eastern U.S. sources more than 1,000 km (600 miles) distant from receptors probably contribute much less acidic deposition than do closer sources.
• Lack ability to measure reliably dry deposition which may be especially important for local source contribution to total deposition.	• Existing models and data analyses can give a qualitative sense of the relationship between sources and receptors.
• Existing atmospheric models cannot predict event variability in deposition but episodes of high acidity may cause much of the acidification.	• SO_2 emissions reductions over a broad area for a long time period may provide essentially proportionate reductions in acid deposition, if all other climatological inputs are held constant.

Although we do not completely understand the science of acid rain, including how quickly the environment would benefit from reduced emissions, the focus of the debate has shifted increasingly away from scientific questions to how to allocate the costs of emissions reduction programs. Because sulfate is the major constituent of acid deposition—but not necessarily the major cause of ecological effects—in eastern North America as well as Europe, advocates of controls have emphasized reducing sulfur dioxide emissions. But analysis of European Air Chemistry Network (EACN) data reveals a doubling of NO_3^- concentrations at most stations between the late 1950s and early 1970s, with some increase in ammonia nitrogen. Moreover, analysis of the data for Swedish stations revealed no further increases in either ion from 1972 to 1984, while $SO_4^=$ concentrations decreased by 30 percent (Rodhe and Rood 1986). As a result, a prudent strategy, given the role of nitrates and ozone, also should focus on limiting NO_x and HC emissions.

The addition of economic factors to the acid rain debate has not meant, however, that the scientific dimension of the problem has decreased in importance. On the contrary, research and development have been given even greater importance because of the controversy over whether further emissions reductions are necessary to cope with acid rain. Both the public and private sectors have channeled an increasing number of scientists and a greater level of funding into acid rain R&D programs.

Federal Acid Rain Research

The previous chapter described the political attractiveness of a research and development approach to acid rain and outlined the dramatic expansion of the federal R&D program. But the acid rain research effort continues to be a major policy concern. Three issues have dominated the debate about the federal government's R&D in the acid rain arena: the timing, breadth, and substance of the program.

The federal government's acid rain research program, commonly known as the National Acid Precipitation Assessment Program (NAPAP), was authorized by the Acid Precipitation Act of 1980. The act established the Interagency Task Force on Acid Precipitation to plan and coordinate the implementation of NAPAP. The task force is chaired jointly by representatives of the National Oceanic and Atmospheric Administration (NOAA), EPA, the Department of Agriculture (USDA), CEQ, the Department of the Interior (DOI), and the Department of Energy (DOE). The Tennessee Valley Authority (TVA) participates in NAPAP activities as well. The other members of the task force who participate in but do not fund NAPAP activities are the Department of Health and Human Services, the Department of Commerce, the National Aeronautics and Space Administration, and the National Science Foundation. The task force also involves the federal government's national laboratories through a consortium consisting of Argonne, Brookhaven, Oak Ridge, and Pacific Northwest National Laboratories, and includes four presidential appointees. NAPAP is structured around a series of task groups organized by scientific areas of research such as natural sources, manmade sources, and aquatic effects. Established to carry out the operational goals of the Interagency Task Force on Acid Precipitation, NAPAP has a mandate to develop a comprehensive research program to synthesize understanding of the problem.

The federal government's funding for acid rain research in constant dollars is illustrated in tables 8 and 9. While the shares for individual agencies have remained relatively stable over time, changes in funding have been dramatic across task groups. In addition, overall funding increased substantially from FY 1982 to FY 1987, accompanied by shifts in emphasis among research areas. For example, research on ecological effects—particularly forestry studies regarding terrestrial effects—and atmospheric processes have received more funding. Support for assessment and policy analysis has been more volatile, reflecting the problems of integrating efforts in that area. Similarly, funding for the control technology task group essentially reflects engineering evaluations of the limestone injection multistage burner (LIMB) technology's potential for reducing SO_2 and NO_x emissions.

NAPAP was to provide preliminary estimates of acid rain impacts by 1985, focusing on the northeastern United States. These estimates were to be made public by September 1986. This initial damage assessment, however, had not been released

TABLE 8
NAPAP Funding of Agencies,
FY 1982–FY 1987

	1982	1983	1984	1985	1986	1987	Total
EPA	9,125	11,093	13,585	31,061	45,061	47,386	157,311
		(0.22)	(0.22)	(1.29)	(0.45)	(0.05)	
DOI	2,110	3,460	4,645	7,556	5,181	5,675	28,627
		(0.64)	(0.34)	(0.63)	(−0.31)	(0.10)	
DOE	2,544	1,998	3,654	7,862	7,018	7,742	30,818
		(−0.22)	(0.83)	(1.15)	(−0.11)	(0.10)	
USDA	1,349	2,874	2,611	8,190	11,336	8,594	34,954
		(1.13)	(−0.09)	(2.14)	(0.38)	(−0.24)	
NOAA	1,900	2,183	2,172	3,733	3,569	3,567	17,124
		(0.15)	(−0.01)	(0.72)	(−0.04)	(−0.01)	
TVA	325	—	—	—	—	344	669
Total	17,353	21,608	26,669	58,403	72,164	73,308	269,505
		(0.25)	(0.23)	(1.19)	(0.24)	(0.02)	

Source: Computed from data provided by National Acid Precipitation Assessment Program, Personal Communication, R. Downing, January 1987.
Note: Figures are given in 1982 constant dollars in thousands. Percent change from the previous year is given in parentheses.

to the larger scientific community and general public by mid-1987. Nonetheless, two additional integrated, policy-related assessments drawing upon the results of the initial NAPAP study were scheduled for 1987 and 1989, according to NAPAP's original operating plan. The 1985 estimates also were to serve as a mechanism for fine-tuning future R&D efforts.

Delays in meeting deadlines have raised serious questions about NAPAP's management as well as ability to provide policy-relevant, scientific information in a timely fashion. Such concerns about the technical quality and progress of NAPAP, as well as future research directions, are not new. For example, in September 1983, an ad hoc external review panel chaired by John Deutch of the Massachusetts Institute of Technology was created to review NAPAP because EPA Administrator William Ruckelshaus wanted a fresh look at the program's adequacy and relevance to policy concerns.[8] The conclusions of the Deutch panel underscore the strengths and weaknesses of the federal government's research effort:

TABLE 9
NAPAP Funding of Task Groups, FY 1982–FY 1987

Task Group	1982	1983	1984	1985	1986	1987	Total
Natural sources	600	679 (0.13)	788 (0.16)	1,040 (0.32)	732 (−0.29)	778 (0.06)	4,6
Manmade sources	1,170	1,310 (0.12)	1,953 (0.49)	2,102 (0.08)	3,575 (0.70)	3,365 (−0.06)	13,4
Atmospheric processes	4,854	5,075 (0.05)	5,945 (0.17)	12,617 (1.12)	13,556 (0.07)	18,525 (0.37)	60,5
Deposition monitoring	3,034	4,277 (0.41)	5,572 (0.30)	7,651 (0.37)	9,121 (0.19)	9,455 (0.04)	39,1
Aquatic effects	3,052	3,262 (0.07)	3,990 (0.22)	14,306 (2.57)	18,279 (0.28)	17,689 (−0.03)	60,5
Terrestrial effects	2,850	4,304 (0.51)	4,192 (−0.03)	14,711 (2.51)	21,697 (0.47)	19,440 (−0.10)	67,1
Effects on materials and cultural resources	428	965 (1.15)	1,390 (0.44)	1,786 (0.28)	1,754 (−0.02)	1,815 (0.03)	8,1
Control technologies	—	—	—	563	548 (−0.03)	278 (−0.51)	1,
Assessment and policy analysis	1,365	1,736 (0.27)	2,840 (0.64)	3,629 (0.28)	2,901 (−0.20)	1,962 (−0.32)	14,
Total	17,353	21,608 (0.25)	26,669 (0.23)	58,403 (1.19)	72,164 (0.24)	73,308 (0.02)	269,

Source: See table 8.
Note: Figures are given in 1982 constant dollars in thousands. Percent change from the previous y is given in parentheses.

On the one hand, the Review Committee was favorably impressed with the progress that NAPAP has made since its establishment in 1981, the manner in which the interagency process is working to coordinate research projects in the participating agencies, and the scope of the present research effort.

On the other hand, the Committee has found several weaknesses in the program which should be repaired. First, insufficient resources are being provided to NAPAP given the enormous breadth and complexity of the technical issues involved. Moreover, the resources are allocated to the participating agencies in a manner which competes with declining agency research budgets and does not give part-time task group leaders authority over the technical program for which they are responsible. Second, the present decentralized interagency management process is not likely to be capable of undertaking several important functions: systematic integration of research results, management of large scale projects, and technical support for policy formulation. Third, insufficient multi-year, indepth studies of the atmospheric consequences of emissions and ecological effects of acid deposition on lakes and streams, watersheds, forests, soils, and biota are underway to provide an adequate basis for verifying "system" models which are needed to meet credibly the 1985 and 1987 NAPAP assessment milestones. Substantial additional resources will be required over time to answer important technical issues. Key areas requiring additional emphasis are integrated assessments, indepth studies of aquatic and terrestrial effects and verification of source-receptor models (Deutch et al. 1983, 1).

Those criticisms—especially NAPAP's inability to meet deadlines for providing integrated assessments necessary to building policy consensus—remain largely justified. As a result, in its 1985 annual report to the president and Congress, NAPAP modified its assessment schedule. The present plan is "to have a first interim assessment in 1987, a second interim assessment in early 1989, and a final assessment in 1990" (Interagency Task Force on Acid Precipitation 1985, 8).

Quite obviously, the pace of NAPAP's operations causes concern among some parties-at-interest. For example, asserting that action should be taken now, environmentalists insist that it is not prudent public policy to wait for further scientific results before deciding on a control program. As the U.S. Office of Technology Assessment has noted (1984, 116), "While such a research plan might produce more accurate analyses than those resulting from a shorter effort, many consider waiting until 1989 for this level of refinement unacceptable." Yet equally strong forces press for delays. Spokesmen for industry and coal miners are especially committed to more deliberate research strategies so as to build on findings from research now being conducted. They argue that shortening the time allotted to R&D will reduce the utility, not to mention the credibility, of those results.

Similarly, the breadth of the federal program is at issue. Some scientists believe that NAPAP's focus on acid deposition resulting from sulfate and nitrate inputs is too restrictive. They argue that multiple pollutants should be considered in

NAPAP's acid rain studies and that greater coordination of research indirectly linked to the acid deposition issue should be a major policy goal (U.S. Office of Technology Assessment 1984).

Finally, some observers of the federal R&D program have serious doubts about its focus. Efforts to expand the R&D program along substantive lines fall into two categories, both of which have found their way into legislative proposals. First, there have been efforts to expand and accelerate research into causes and effects. This thrust has emphasized more funding for mitigation methods, especially liming programs for neutralizing acidity in areas already affected, including restoring aquatic or forest buffering capacity.[9] Second, and perhaps more significant, some participants in the debate have begun to press for more aggressive R&D for developing and demonstrating technology for the use of cleaner coal. This attempt to reduce acid rain's precursor pollutants at the precombustion or combustion stages focuses on coal science, process science and engineering, and engineering and development programs. Such a program would actually sponsor technological demonstrations of available alternatives for coal combustion (Blodgett, Parker, and Backiel 1984).

3
ACID RAIN CONTROL TECHNOLOGIES AND MITIGATION STRATEGIES

Better generating technologies are in fact already available or on the engineering horizon.... None of these solutions is environmentally perfect, and many have yet to be proved in commercial operation. Some will surely not live up to environmental or economic expectations when they are tried. But if we confront acid rain's uncertain science honestly, we must recognize that there is no real choice but to seek out and experiment aggressively with these new alternatives.

—*Peter Huber (1984)*

ONE OF THE clearest manifestations of the distinction between the acid rain problem as an air quality issue and as an emerging policy arena is seen in control technologies. On the one hand, various technological options based on environmental engineering have been developed as direct responses to the demands of the Clean Air Act. The CAA's technology-forcing provisions brought a host of coal cleanup alternatives under consideration very early in the clean air debate. These alternatives can be used prior, during, or after the actual combustion of coal as a fuel. Thus, one of the earliest efforts to formulate a framework for environmental policy discussed in detail the availability of various technological strategies to remove pollutants to affect their concentration or to mix techniques to influence the dispersion of sulfur oxides from coal combustion in the atmosphere (Garvey 1972).

But it is also true that the unique characteristics of the acid rain problem have broadened the menu of technological choices in recent years. Nowhere is this more obvious than in the 1986 report by President Reagan's special representative on acid rain, Drew Lewis, and his Canadian counterpart, William Davis, which recommended a joint government-industry program to develop over a five-year period ways to burn coal cleanly (Sun 1986).

Expanding the range of control technologies has, at least in theory, the advantages of reducing the constraints on feasible policy options. Unfortunately, the availability of more technological choices also complicates the process of consensus-building. This is because there is typically substantial disagreement about whether a particular technological option is actually available, has near-term potential, or merely is theoretically possible (Kash and Rycroft 1984; 1985). This chapter examines the variety of engineering control technologies that might be appropriate for addressing the acid rain problem according to these three categories of development (see table 10). In addition, we offer a brief overview of liming as a mitigation strategy for renovating and protecting selected aquatic ecosystems.

Available Technologies

A technology is "available," or "off-the-shelf," if reliable estimates can be made about what it will cost and how long it will

TABLE 10
Available, Potential, and Possible Technologies for Controlling Acid Deposition

Available	Potential	Possible
Physical coal cleaning	Chemical coal cleaning	Other electricity fuel generation methods
Oil desulfurization	Limestone injection multistage burner	Energy conservation
Wet flue gas desulfurization	Atmospheric fluidized bed combustion	
Dry flue gas desulfurization	Pressurized fluidized bed combustion	
Regenerable flue gas desulfurization	Integrated gasification combined-cycle technology	
Staged combustion	Magnetohydrodynamics	
Flue gas recirculation	Electron beam irradiation	
Dual-register burner		
Low excess air		
Low nitrogen oxide burner		
Flue gas treatment		

take to construct or carry through to conclusion. Unlike the debate of a decade ago, current debate about available technologies such as flue gas desulfurization (FGD) systems does not focus primarily on performance. This is not to say that all uncertainties about reliability and maintenance problems or about how well a particular design option will operate under field conditions have been fully resolved. But the focus of debate has shifted largely to issues such as who will pay and who will derive benefits, or what is the likely distribution of risks from placing a particular pollution control technology in a specific location. The debate also includes management concerns such as how to go about linking specific pollution control technologies to institutional mechanisms. In the area of acid rain control, several technologies have been available for some time (see U.S.-Canada Work Group 3B 1982).

Physical Coal Cleaning

The physical cleaning of coal is the major precombustion technology now in use for reducing the acid rain problem. Physical coal cleaning (PCC) approaches rely on the fact that coal has a lower specific gravity than the impurities such as sulfur associated with it. Thus, the two can be separated by washing them in a water medium that allows the crushed coal to float and the impurities to sink (see figure 6). A range of PCC alternatives have been developed, with the key factor being the density or size of the particles to be treated. At least seven such applications have been identified (Green 1984). Their potential for reducing sulfur emissions depends on the initial sulfur level in the raw coal and the ratio of organic to inorganic sulfur present (U.S. Office of Technology Assessment 1983). This is because physical coal cleaning removes only inorganic impurities consisting mainly of pyrite. Taking these factors into account, the variation in sulfur removal among PCC technologies has been estimated at between 20 and 35 percent (Parker and Trumbule 1983). This could translate into emissions reductions on the order of 8 to 33 percent (Gould 1985; U.S. Office of Technology Assessment 1984). Largely because of these relatively low removal efficiencies, physical coal cleaning has been applied only to an estimated half of the high-sulfur steam coal used by the utility industry (Yeager 1984).

FIGURE 6
Process Flow for a Typical Physical Coal Cleaning Plant

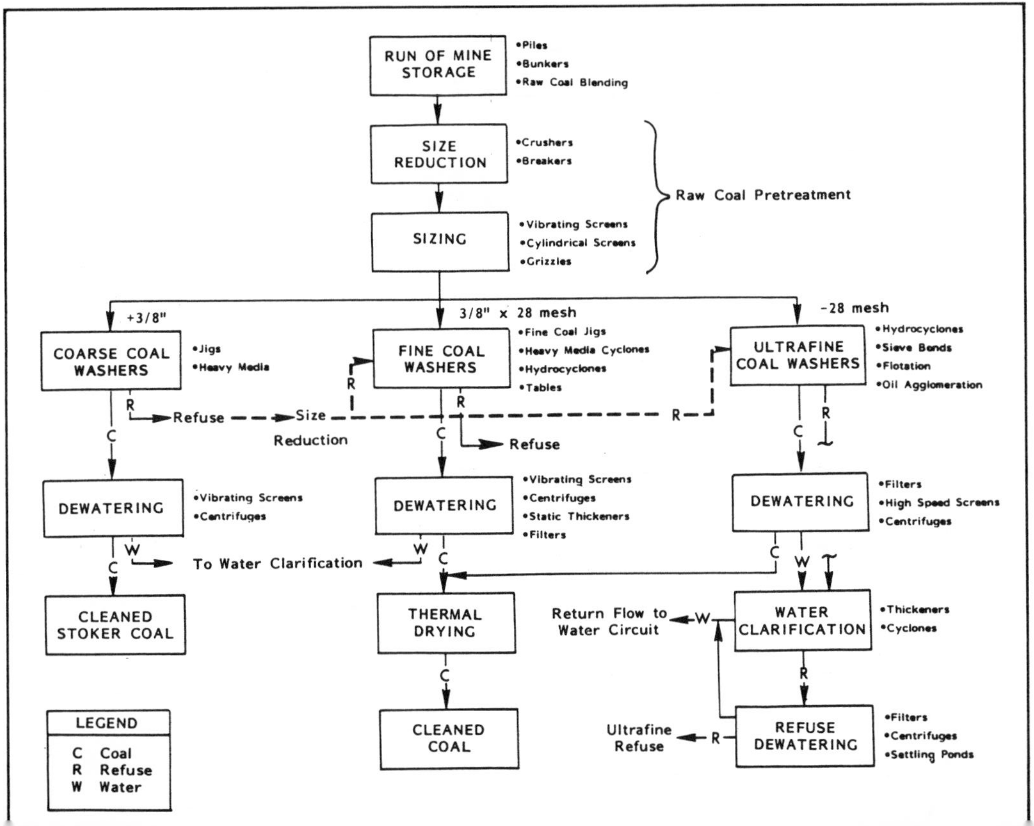

Yet coal cleaning is attractive for a number of reasons. First, it requires only minimal power plant modification. Second, it reduces transportation costs and may improve the combustion process. And third, physical coal cleaning may be used in conjunction with other control options to reduce the burden on combustion or postcombustion cleanup (Parker and Kaufman 1985; Electric Power Research Institute 1983). It must be noted, however, that coal cleaning does produce a substantial amount of solid waste.

Cost has been a controversial aspect of the debate over coal cleaning because of the range of technologies available and the variety of removal levels associated with different types of coal (Hutton and Gould 1982). In short, the economics appear to be site-specific (Parker and Trumbule 1983). However, one generalization can be made. For lower sulfur coals, the cost of physical cleaning processes increases rapidly with decreases in sulfur content. The cost-effectiveness of coal cleaning technologies improves as the sulfur content of the coal increases, as coal prices rise, and as utilities try to increase generating capacity without the need for major capital investments. As a result, this is where most of the effort to date has focused and physical coal cleaning has been identified by the Office of Science and Technology Policy (1983) as one of the first and least expensive steps that could be taken to reduce sulfur emissions.

Fuel Oil Desulfurization

Fuel oil desulfurization is a well-developed and widely used technology. The residual sulfur after-treatment of fuel oil to reduce its sulfur content typically ranges between 0.2 to 0.5 percent, although a higher degree of sulfur reduction is technically achievable. As in coal cleaning, the costs for reducing the sulfur content of fuel oil increase with the level of sulfur removal (U.S.-Canada Work Group 3B 1982).

Wet Flue Gas Desulfurization

The dominant policy response to achieve air quality improvements has been the mandatory implementation of wet flue gas desulfurization technologies (see North Atlantic Treaty Organization 1982; Torstrick et al. 1980). These so-called wet scrubbers are the postcombustion option that has been designated as the best available technology in the clean air debate. Because of their over 90 percent removal of SO_2 emissions, wet FGD alternatives represent the classic technological fix in this policy arena. Table 11 reveals that there are over 100 of these wet FGD systems in operation at electric utility power plants in the United States today, and approximately an additional seventy are under construction or on contract.

Wet FGD uses either limestone or lime suspended in water as a medium through which the stack gases of a facility are passed. In this process, the flue gas is sprayed with a slurry made up of water and either lime or limestone, at which time the SO_2 reacts chemically and is removed in the form of a wet sludge (see figure 7). These systems have sometimes proved unreliable and economically inefficient. Nevertheless improvements in reliability and cost-effectiveness have been made, especially in the understanding of system chemistry (Tearney, Froelich, and Graves 1984). The fact that wet FGD technology has been the most widely employed alternative has meant that it enjoys the advantages of being more available, more versatile, and proven through longer experience (International Energy Agency 1985).

Costs for wet FGD vary widely, but there is some consensus that capital investment in this approach is something on the order of five to six times as great as for physical coal cleaning, with operating expense perhaps four times as great (Parker and Trumbule 1983). Moreover, it appears retrofitting older facilities with wet scrubbers entails severe financial liabilities. Capital costs in this instance may run 10 to 40 percent more than for installing wet scrubbers during the construction of new power plants (Electric Power Research Institute 1983).

In addition, a typical 1,000 megawatt (MW) plant burning 3.5 percent sulfur coal will produce about 225,000 tons of solid waste per year. Disposal of that volume of toxic sludge is expensive and dangerous throughout the life of a coal-fired power plant equipped with a wet FGD system. When these economic factors are combined with the significant environmental costs associated with generating large quantities of toxic sludge, many parties-at-interest in the air quality debate have argued for alternative FGD approaches. The result has been the development of the dry FGD technology.

Existing and Projected Flue Gas Desulfurization Systems in U.S. Utilities

	Total		Operational		Under Construction		Contract Awarded		Planned	
	N	MW	N	MW	N	MW	N	MW	N	MW
Processes for Saleable Products										
Aqueous carbonate/spray drying	1	100	0	0	1	100	0	0	0	0
Citrate	1	60	1	60	0	0	0	0	0	0
Lime	1	65	0	0	1	65	0	0	0	0
Limestone	1	166	0	0	1	166	0	0	0	0
Lime/limestone	1	475	0	0	0	0	1	475	0	0
Magnesium Oxide	3	724	0	0	3	724	0	0	0	0
Wellman Lord	8	2,074	7	1540	1	534	0	0	0	0
Subtotal	16	3,664	8	1,600	7	1,589	1	475	0	0
Processes for Throwaway Products										
Dual alkali	5	2,023	3	1,181	1	421	1	421	0	0
Lime	35	17,173	22	9,978	6	3,155	0	0	7	4,940
Limestone	79	36,500	39	13,961	18	8,454	10	6,750	12	7,335
Limestone/alkaline flyash	2	1,480	2	1,480	0	0	0	0	0	0
Lime/alkaline flyash	11	4,013	9	2,613	2	1,400	0	0	0	0
Lime/limestone	2	20	2	20	0	0	0	0	0	0
Lime/sodium carbonate	1	100	0	0	1	100	0	0	0	0
Lime/spray drying	11	3,467	1	110	7	2,338	3	1,019	0	0
Process not selected	10	5,315	0	0	0	0	0	0	10	5,315
Sodium carbonate	9	3,155	5	1,255	0	0	0	0	4	1,900
Sodium carbonate/spray drying	1	440	1	440	0	0	0	0	0	0
Subtotal	166	73,686	84	30,138	35	15,868	14	8,109	33	19,490
Total	182	77,350	92	31,738	42	17,457	15	8,665	33	19,490

Source: North Atlantic Treaty Organization (1982).
Note: As of September 1981. Forty units were undecided and are not included in this table.

FIGURE 7
Limestone Wet Scrubbing FGD Process

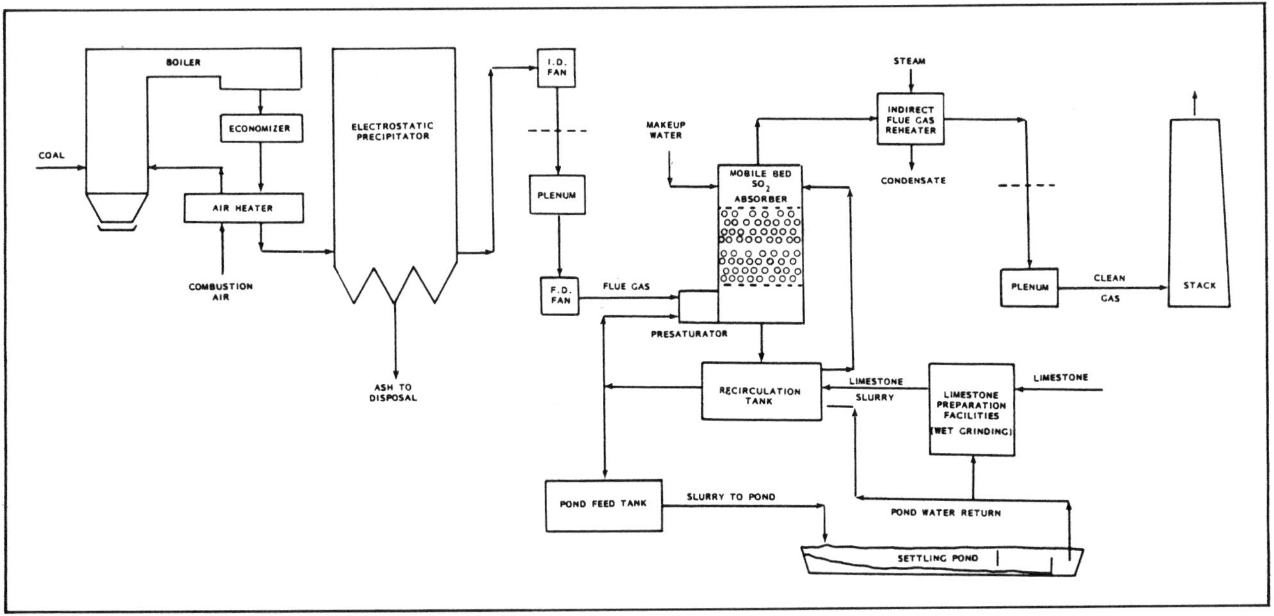

Source: Torstrick et al. (1980).

Dry Flue Gas Desulfurization

Dry FGD is a recent addition to the list of available emissions control options. The first full-scale successful operation of this technology took place in December 1980. Use of dry FGD systems has expanded rapidly since then and U.S. electric utilities had contracted for eleven installations using dry FGD systems with low-sulfur coal as of mid-1981 (U.S.-Canada Work Group 3B 1982, 110). This is apparently because the dry FGD process has a number of advantages over wet scrubbers. Although the operational experience with dry scrubbers is not as extensive as with wet FGD, dry alternatives seem to enjoy slightly lower capital and operating costs than wet technologies. And, because they employ either a lime slurry or soda ash solution injected into a spray dryer, they produce a dry waste product that, while greater in volume, is easier to control and recycle than wet sludge (see figure 8). This technology also requires less manpower, energy, and water. The low water requirement is extremely important in the West. It also is generally assumed that dry scrubbers will be more dependable and simpler to operate than wet scrubbers (U.S. Office of Technology Assessment 1984). For these reasons, the dry FGD technology is viewed by many experts as more promising for retrofit purposes.

However, there are drawbacks. Dry scrubbers have lower SO_2 removal levels than wet systems; while wet FGD removes more than 90 percent of the sulfur content in coal, the removal efficiency for dry systems ranges from 50 to 90 percent depending upon the sulfur content (U.S.-Canada Work Group 3B 1982, 113). A relatively high stoichiometry is necessary for the dry FGD approach to achieve a sufficient degree of sulfur removal from high-sulfur coals. In addition, dry systems are economically less efficient for high-sulfur coals. This has limited dry FGD systems to low-sulfur western coals largely because the operating costs of dry systems—the reagent, lime, is expensive, and more energy is required to operate dry systems—erode their suitability for high-sulfur coals. These are serious limitations, given the current emissions limits for new plants. However, they are less of a constraint if retrofit applications are required to reduce emissions from pre-1971 plants.

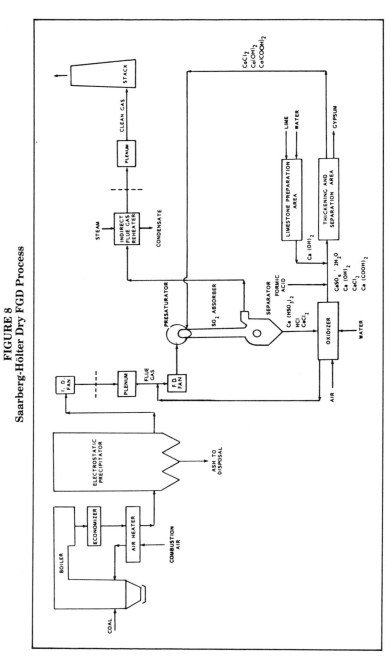

FIGURE 8
Saarberg-Hölter Dry FGD Process

Source: Torstrick et al. (1980).

Regenerable Flue Gas Desulfurization

A third FGD option represents a slight variation on the wet and dry technologies discussed above. In an attempt to overcome the liabilities that are associated with both wet and dry waste management, several different types of flue gas desulfurization technologies have been developed to reclaim the SO_2 and use chemicals such as sodium sulfite to produce marketable products—sulfuric acid or elemental sulfur (see figure 9). Unfortunately, because these regenerable technologies involve substantially higher capital investments, they have received only limited use. In addition, the need to find markets for the sulfur byproducts also has lessened utilities' interest in regenerable FGD systems. For example, in 1982, there were only "eight operational regenerable FGD systems controlling about 1,600 MW of electric generating capacity" in the United States (U.S.-Canada Work Group 3B 1982, 112). Although regenerable FGD systems such as the Wellman-Lord process, which is used by New Mexico Public Service and the Northern Illinois Public Service Company, can achieve sulfur removal efficiencies in excess of 90 percent, the relatively high cost of such systems is a major drawback to their widespread use. In fact, these capital constraints have proven to be formidable barriers to the dissemination of technologies that once were regarded as among the most promising long-term solutions to many coal combustion problems (Lunt and MacKenzie 1984).

Thus far, the focus has been entirely on available sulfur-removal options. At least five additional technologies are available for controlling nitrogen oxides emissions. The first four of these involve modifications made during combustion. The fifth is a postcombustion alternative.

Staged Combustion

The first nitrogen oxides control technology, staged combustion, was applied in 1962. Often referred to as two-stage combustion, this process involved supplying less air than is necessary to burn the coal completely in the first stage of combustion. This releases nitrogen gas, which cannot be oxidized. Then the second stage burns the remaining fuel. The process achieves NO_x reductions on the order of 30 to 50 percent and is relatively

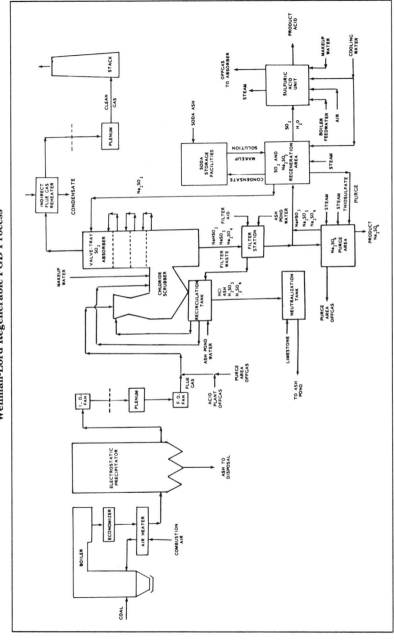

FIGURE 9
Wellman-Lord Regenerable FGD Process

Source: Torstrick et al. (1980).

inexpensive. But staging leads to high levels of corrosion and requires more maintenance. This was the case especially when the technique was applied to pulverized coal (Barsin 1984).

Flue Gas Recirculation

At the same time that staged combustion was being applied, a second approach was developed. Flue gas recirculation technology, as the name implies, merely substitutes a recirculation of the flue gases for the second stage of two-stage combustion. Slightly higher NO_x reductions up to the 60 percent removal level can be achieved for certain types of coals. However, the installation costs of this alternative have proven to be higher and its fuel efficiency lower than for its staging technology (Parker and Trumbule 1983).

Dual-register Burner

The response to these cost and fuel efficiency problems led to the commercialization of dual-register burners in 1972. This technology involves a burner designed to control simultaneously peak temperatures and oxygen availability. Dual-register burners have accomplished levels of NO_x removal in the 50 to 60 percent range. This level of emissions reduction is comparable to those of the earlier technologies. In addition, dual-register burners can achieve those reductions with pulverized coal.

Low Excess Air

By the late 1970s work had begun on additional combustion modifications with the goal of increasing flexibility and reducing costs. One result is the low excess air (LEA) technology. LEA simply reduces the amount of combustion air available and thus reduces the formation of thermal and fuel NO_x. This can be accomplished by changes in combustion operating practices and involves no hardware (U.S. Office of Technology Assessment 1984). Total NO_x reductions of 15 to 38 percent are attained in this highly inexpensive manner.

Low Nitrogen Oxide Burners

Low NO_x burners are a second-generation staged combustion technology (Maulbetsch et al. 1986). Like earlier ap-

FIGURE 10
Typical Low NO$_x$ Burner

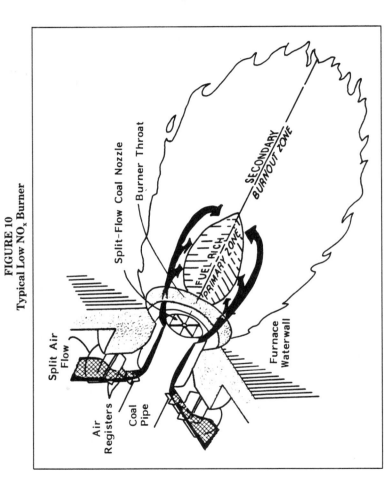

Source: Maulbetsch, McElroy, and Eskinazi (1986).

proaches, they delay the combustion process, but this generation of burners has been designed to be compatible with all four of the basic furnace types now in use in utilities (see figure 10). Therefore, low NO_x burners appear especially attractive for retrofit purposes. NO_x reduction levels of 40 to 60 percent appear feasible at relatively low cost (Electric Power Research Institute 1983).

Flue Gas Treatment

The final NO_x reduction technology, flue gas treatment (FGT), is, as noted above, a postcombustion alternative. There are at least 50 FGT processes available, with the key distinction being whether a catalyst is used. The most promising option, according to most experts, is a process called selective catalytic reduction (SCR), which mixes the flue gases with ammonia and then passes this mixture over a catalyst such as copper oxide. The consequence is a reaction of ammonia and NO_x to form water and nitrogen gas. As much as 90 percent NO_x removal has been accomplished with SCR, but it comes with very high capital and operating costs—as much as ten times the cost of combustion modification approaches. In addition, this technology has problems associated with the disposal of catalysts, and equipment corrosion and damage have been identified as possible barriers. A noncatalytic approach, selective noncatalytic reduction (SNR) is similar to SCR, but SNR depends entirely on the high temperature of the gases to facilitate the reaction with ammonia.

What generalizations can be made about this menu of available technologies for controlling SO_2 and NO_x? First of all, the dominance of throwaway scrubber technology is obvious. That dominance could continue into the late 1990s. But it is highly controversial among both energy and environmental advocates.

A nontechnological alternative—changing fuels—is attractive. But fuel switching may not be politically feasible. Because of its socioeconomic consequences, particularly shifts among regions in their coal market shares if this option is permitted, switching from high- to low-sulfur coal would create the risk of substantial losses in production and employment in the coal-producing regions of the midwestern and eastern United States.

Moreover, expanding the choice of technologies to include options now only potentially or theoretically available is complicated by a number of factors. Most significantly, there is great uncertainty regarding the timing and urgency of control policy. If society decides to undertake a rapid "crash program" of emissions reduction, as advocated by increasing numbers of parties-at-interest, then the technological choices will be largely limited to currently available alternatives. The more deliberate approach currently being followed could mean an expanded set of technological choices. In particular, more time might make it possible not only to improve SO_2 and NO_x reduction technologies, but also to develop and deploy combined SO_2 and NO_x control alternatives.

Potential Technologies

A technology can be characterized as potential if it is feasible, but there is debate about what it will cost and how long it will take to complete its penetration of commercial markets. Unlike available alternatives, potential options raise questions not only about impact and management, but also about performance. The following major technologies fall into the potential category.

Chemical Coal Cleaning

Improving the SO_2 reduction capabilities of coal cleaning requires moving from physical to chemical methods (see U.S.-Canada Work Group 3B 1982). Chemical coal cleaning (CCC) involves the use of chemical reagents to remove not only inorganic but also organic impurities. Among the processes now under development, the most promising involve froth flotation of water and the use of fine magnetites suspended in a cycloning vessel by centrifugal force. In either case, the higher sulfur removal levels appear to be offset by higher costs.

Development of these processes is in a very early stage (Parker and Kaufman 1985). No commercial CCC process is in operation in the United States; commercial viability is estimated to be at least five to ten years away. Their economic competitiveness is highly uncertain because of high costs, high energy losses, and potential engineering difficulties in moving from the pilot stage to commercial development (U.S. Office of Technology Assessment 1984). Commercialization appears to depend on achieving a sufficiently high reduction rate to eliminate the need for further control modifications. While some laboratory tests show sulfur reduction rates in excess of 90 percent, on the whole, the prospects for the CCC alternative remain very much in doubt.

Limestone Injection Multistage Burner

In stark contrast to the doubts about chemical cleaning, the limestone injection multistage burner (LIMB) technology seems a promising alternative. LIMB involves the use of staged combustion techniques combined with limestone injection. In this process, SO_2 reacts with the limestone to create a solid material, calcium sulfate. The basic principles are hardly new; efforts to develop this process go back to the 1960s. However, a better understanding of the underlying chemical processes have dramatically improved the potential of the LIMB alternative (Poundstone and Rubin 1985).

Two characteristics make LIMB highly attractive. First, it reduces both SO_2 and NO_x. Although removal rates are highly uncertain, some estimates project between 50 and 70 percent reduction in emissions by the time of commercial application (Parker and Trumbule 1983). Japanese experience with the LIMB technology at the pilot level and similar experience in West Germany have demonstrated removal rates of about 50 percent efficiency. As a result, there is a substantial consensus within the engineering community that removal at the 40 to 60 percent level is a reasonable expectation. While this is only about half the rate for wet FGD, the costs of LIMB appear to be only about 10 to 25 percent that of scrubbers (Electric Power Research Institute 1983). A second perceived advantage of LIMB is the flexibility it promises for retrofitting. Retrofit experience in West Germany indicates that LIMB is both easily and inexpensively applied to existing coal-fired facilities. Adding to its attractiveness is the fact that LIMB may provide badly needed renovation of oil-fired facilities as well (Energy Research and Advisory Board 1985).

Atmospheric Fluidized Bed Combustion

Another promising combined SO_2 and NO_x control technology is atmospheric fluidized bed combustion (AFBC). This alternative injects powdered coal and limestone onto a "fluidized bed" raised by air pressure from below. As with LIMB, solid calcium sulfate is then produced, but unlike LIMB, AFBC relies on lower combustion temperatures to keep down nitrogen oxide formation. There are several advantages to this approach, including its higher emissions reduction capability. The AFBC technology may achieve as much as 90 percent removal of SO_2. It is also less expensive, is more energy efficient, and can be used in facilities with smaller boilers. Moreover, AFBC can burn both high- and low-sulfur coal and produces a dry waste product that can be disposed of easily. The retrofit capability for AFBC is now somewhat uncertain. But there are potential retrofit advantages in the fact that AFBC can be applied to a range of facility sizes and can burn a wide range of fuels of different qualities. The major technical issues, at present, have to do primarily with problems of AFBC scale-up that remain to be resolved (Squires et al. 1985).

Pressurized Fluidized Bed Combustion

Whereas both LIMB and AFBC are commonly assumed to be commercially available by the early 1990s if not before, two additional potential control options most likely will not come "on line" until sometime in the mid-1990s. The first of these is the pressurized fluidized bed combustion (PFBC) technology. As the term indicates, PFBC operates at higher pressures than AFBC technology. Other key characteristics are the smaller size of PFBC units, reduced costs, less waste, higher efficiency, and more NO_x removal capability. On the other hand, PFBC has lower SO_2 removal rates, on the order of 50 percent. In addition, there are a host of unresolved questions having to do with adverse effects of PFBC on the boiler (Markowsky 1984).

Integrated Gasification Combined-cycle Technology

The second potential option for the mid-1990s is integrated gasification combined-cycle (IGCC) technology. First the coal is gasified. Then the clean gas is burned and the exit gas is used to power a steam turbine, which is why the technology is called a "combined cycle." A gasification facility of this type offers some special advantages for acid rain control. Sulfur dioxide removal may be as high as 95 percent, NO_x emissions reduction rates also appear promising, and the IGCC adds the benefit of coproduction of electricity (Crawford 1985). Coal gasification technology has been commercialized abroad. A demonstration IGCC facility near Barstow, California, has been operating since May 1984, and so far has had promising results (Spencer et al. 1986). There is uncertainty, however, about the competitiveness of gasification with conventional fossil fuels (Holt and O'Shea 1984).

Magnetohydrodynamics

The use of magnetohydrodynamic (MHD) combustion to produce electricity holds the potential for virtually eliminating SO_2 emissions and substantially reducing NO_x emissions. This technology relies on producing electricity directly from thermal energy. This is done by channeling an electrically conducting gas through a magnetic field, thus creating a voltage drop in the stream of gas (Parker and Trumbule 1983). Conventional wisdom among energy researchers suggests that while MHD is technically feasible, it must overcome substantial problems in materials development and scale-up. Commercial availability is not expected until some time after the year 2000, at the earliest.

Electron Beam Irradiation

Also technically feasible, but assumed to be relatively long-term in its potential for commercial application, is electron beam irradiation (EBI). This approach uses electron beams to dissociate NO_x and SO_2. EBI then subjects those pollutants to ammonia to create ammonium sulfate and ammonium bisulfate. Japanese R&D work in this area at the pilot plant level is under way (Parker and Trumbule 1983).

Theoretically Possible Technologies

Theoretically possible technologies feature not only debates about the range of performance, impact, and management questions, but also controversy regarding whether such an activity can be done at all. The only consensus underpinning theoretically possible technologies is that they cannot be carried out within any predictable time frame or with any predictable costs, benefits, or risks.

Other Electricity Generation Methods

The substitution of alternative electricity generation technologies is one theoretical approach to reducing both SO_2 and NO_x. Most of the focus has been on increasing the use of new nuclear or solar options to replace coal-fired capability. Of course, there is no purely technical barrier to the replacement of coal-fired electricity with either available or potential nuclear, oil, or gas technologies. Yet large-scale substitution of this kind in the near future simply does not seem likely. This reflects the collapse of the U.S. nuclear power industry and significant reserve-resource questions about the future of petroleum and natural gas. The introduction of theoretically possible options, such as nuclear fusion or any one of a half-dozen solar technologies including ultra-high temperature concentrators that are in the engineering demonstration or early commercialization stage, could dramatically alter the acid rain calculus. Such breakthroughs, however, are not likely to happen soon.

Energy Conservation Technology

If substituting new technologies is a theoretically possible approach to acid rain control from the supply side, then energy conservation may hold equal promise through management of demand. However, as is not true of some of the new technologies, there are fundamental obstacles to altering the acid rain problem with off-the-shelf conservation approaches. First, the United States has already achieved substantial energy conservation gains since 1973. Many of these gains have been achieved by the utilities and industry. The temporary oil glut may create greater willingness to increase consumption in the near term. However, the magnitude of prior conservation gains and their impact on existing capital stock as well as behavior patterns probably means that just as consumption will not go up significantly neither can conservation be taken much further (Regens 1985b).

To be sure, improvements are likely in engineering design

and system operation in the future. But truly significant changes in the patterns of SO_2 and NO_x emissions projected in figure 11 will occur only if breakthroughs take place in areas such as advanced materials for use in thermal energy storage systems.[1] Moreover, long-term development of new energy production and utilization technologies also has been vulnerable to budgetary uncertainty and sharply reduced funding. This is particularly true of the federal government's solar and conservation R&D programs, which have suffered from a unique combination of technological obstacles, political turbulence, and ideological opposition (Frankel 1986).

FIGURE 11
U.S. Sulfur Dioxide and Nitrogen Oxides Emissions Trends, 1900–2030

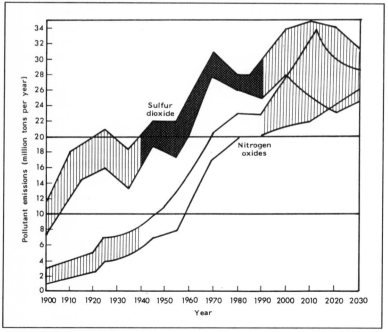

Source: U.S. Office of Technology Assessment (1984).
Note: Pre-1940 estimates and post-1990 projections are highly uncertain. Projections of future emissions incorporate a wide range of assumptions about future economic growth, energy mix, and retirement of existing facilities; they assume no change in current air pollution laws and regulations.

LIMING

The application of alkaline material, generally referred to as liming, to increase the acid neutralizing capacity of sensitive receptors has been proposed as a potential mitigation strategy for dealing with the acid rain problem. What, then, is liming's potential as a mitigative measure? In practice, the utility of adding neutralizing or buffering materials is limited to specific circumstances. For example, liming drinking water systems with low pH levels to control corrosion is a relatively low-cost and standard practice (Hudson and Gilcreas 1976). The final report on impacts of acidification prepared by U.S.-Canada Work Group 1 concluded (1982, 9–11): "Such activities are commonly called 'liming' and they are limited in practice to correcting to some degree adverse effects on aquatic, terrestrial and drinking water systems. It is not possible to use liming to mitigate effects on materials, visibility, and adverse health effects resulting from direct inhalation of airborne pollutants."

Liming appears to be most useful as a technique for increasing the acid neutralizing capacity of surface waters containing sport or commercial fisheries affected or threatened by acidification (see Britt and Fraser 1983; Fraser et al. 1982; Bengtsson et al. 1980). Sweden, for instance, has conducted an extensive lake and stream liming program and experimental efforts are under way in Scotland. The federal government under NAPAP, the Electric Power Research Institute, and Living Lakes, a nonprofit corporation, currently are conducting liming efforts in the United States to investigate the efficacy of such activities for restoring and protecting selected aquatic resources. Theoretically, terrestrial liming, both of watersheds and forests, may neutralize an acidifying pollutant. However, its environmental effects, both good and bad, are less well defined.

IMPLICATIONS FOR POLICYMAKING

Any attempt to characterize our list of potential acid rain control technologies must include the observation made by DOE's Energy Research Advisory Board that "progress is being made and that federal co-funding of demonstration facilities would expedite dimunition of acidity of rain" (Abelson 1985, 819).

Current efforts to make these potential technologies practically available are hampered by the Reagan administration's policy decision to separate research from development. The administration sees federally supported research as an investment in our future, but it believes that most technological development is better suited to the private sector.

The implications of this strategy for acid rain control are far-reaching. Budgets for energy development, demonstration, and commercialization have been dramatically reduced over the last seven years. This has been true for fossil fuels in general and for coal in particular. In fact, the Reagan administration has been reluctant to spend the funds already allocated for clean coal research and development. Private industry has taken key initiatives to make up for this shortfall in federal support. Most of the potential technologies discussed above have benefited from funding provided by corporations and associations such as the Electric Power Research Institute. While government agencies such as DOE, EPA, and TVA have been engaged in a range of cooperative federal-private efforts, without a concerted national program supported by significant federal funding, rapid development of succeeding generations of technological controls may be severely hampered.

Reacting to this potential problem, Congress added to the 1986 budget a Clean Coal Technology Program, in addition to the DOE's $700 million long-range coal research program covering the period FY 1986 to FY 1991. The program's budget is $398 million, to be used over three years. Funding levels are authorized starting at $99.4 million in FY 1986 and increasing to $149.1 million each year for FY 1987 and FY 1988. DOE is to solicit proposals from the private sector for a set of cost-shared, commercial-scale demonstrations of clean coal technologies. These are to include advanced coal cleaning techniques, alternative combustion technologies, preparation of clean coal-based fuels, and postcombustion cleanup systems (American Association for the Advancement of Science 1986; Office of Management and Budget 1986). When this response is added to the joint government-industry program recommendation made by the special envoys of the United States and Canada, it is clear that key participants in the acid rain debate have become mobilized around a more aggressive strategy of technological innovation. Nevertheless, given the budgetary uncertainties generated by the Balanced Budget and Emergency Deficit Control Act of 1985 (Gramm-Rudman-Hollings), any large-scale programs may be extremely hard to legitimize because of their expense.

A strong feeling of déjà vu surrounds the current discussion of acid rain control technologies. Promising technologies always seem to be just over the horizon, if only more funding and expertise could be made available. For example, in the early 1970s, the first comprehensive analysis of energy and environmental policy, conducted by the Energy Policy Project of the Ford Foundation (1974, 193) noted:

Meeting air quality standards will require a wide-spread commitment to the use of scrubbers. Variances and delays in compliance with the standards may be necessary in some cases, because of the lead times for building and installing scrubber systems. In our view, if such variances are needed, they should be coupled with a firm commitment to install scrubbers at the earliest feasible date.

For the longer term, there are several promising clean coal technologies on the horizon to remove the sulfur and ash from the coal before or during combustion. Options such as solvent refining of coal, coal gasification, and fluidized bed combustion may be commercially available, beginning in the 1980s, if adequate funds are devoted to their development.

It is interesting to compare that optimistic 1974 statement with the following assessment of DOE's emission control program, provided a full decade later:

Within the Department's Fossil Energy R&D program, a high degree of emphasis is placed on those combustion and advanced power cycle technologies that offer the potential for meeting or exceeding current NSPS standards in an economical manner. DOE energy projections indicate a significant opportunity around the year 2000 for the introduction of a variety of advanced technologies. This timing is consistent with the potential applicability of long-term technological advances expected to emerge from the DOE fossil energy R&D program (Mares 1983, 240).

Yet we have not managed to add a single new "technological fix" to our coal cleanup arsenal since the issue was placed on the nation's policy agenda by the energy crisis of the 1970s. Clearly something has gone wrong.

One recent analysis of the development of new energy technologies has argued that "our commercialization process is in trouble" for at least four reasons (Lefevre 1985). First, agencies often engage in the over-refinement of technology. One key to our presumed failure to innovate in the energy sector may be the tendency of parent mission agencies to favor technically sophisticated systems, delay their release, and interpret commercialization largely in terms of hardware considerations. A second factor is the loss of focus during the commercial demonstration of programs. Over time, technological development efforts may become skewed by new policy goals or priorities, or changing economic or political conditions. Premature imposition of market criteria also can be a problem. Fluctuating market conditions may affect the attractiveness of technologies, and there may be insufficient commitment to maintaining programs under conditions of economic instability. Finally, the fluctuating impact of tax incentive programs can hinder federal R&D commercialization. Using tax incentives to encourage innovation may be too indirect to have the desired effect.

Each of these factors appears to have influenced the evolution of coal pollution abatement technologies. More stability and comprehensiveness in developing and deploying emission control technologies require a coherence in energy and environmental policy that has so far been lacking. Unfortunately, the current acid rain debate apparently offers little promise of such coherence. Instead, the clash between the utilities and environmental groups over the pace at which new control options should be introduced underscores how far technological commercialization has been politicized. Typical of this conflict is the following debate over coal gasification:

"We're right on the cusp of getting a new kind of pollution control technology that is much simpler, much more cost-effective, much more efficient," said [Robert A.] Beck of the Edison Electric Institute. "But we need about another five or six years to get those things to the point of being commercial."

Environmentalists see this as a stalling tactic. "The argument is similar to the ones made by the utilities in the 1960s when they wanted to put up tall stacks and said, 'Give us some time to develop controls,'" [David G.] Hawkins [of the Natural Resources Defense Council] said. "You see what they've done with that argument. Now that they have the controls, they say, 'Don't make us use those controls but give us time to develop new technologies [to burn coal cleanly].'"

'It's not a stall," responded Charles W. Linderman, the electric institute's fossil fuels program manager. "I think the industry has shown good faith" in developing the technologies, which he insisted are not designed only for environmental purposes—"although there will be a real strong environmental bonus"—but to increase the efficiency of coal-burning boilers (Stanfield 1985, 2368).

Resolving this kind of controversy requires coming to grips with not only the scientific uncertainties outlined in the previous chapter, but also economic and political issues as well.

4
THE ECONOMIC DIMENSION: BENEFITS, COSTS, AND FINANCING OPTIONS

How much are the fish worth in those 170 lakes that account for 4 percent of the lake area of New York? And does it make sense to spend billions of dollars controlling emissions from sources in Ohio and elsewhere if you're talking about a very marginal volume of dollar value, either in recreational terms or in commercial terms?

—*David Stockman (1980)*

IN A SITUATION in which science is not definitive and technology only part of the solution, it is no surprise that the acid rain debate has turned increasingly to economic questions. Today, the key questions from a regulatory policymaking standpoint appear to be: (1) the magnitude of environmental benefits and control costs; and (2) how to allocate the costs of emissions reduction programs. Because sulfate is the major constituent of acid deposition in eastern North America as well as in Europe, advocates of controls have emphasized reducing SO_2 emissions. Yet the very complexity of the phenomenon—especially because distant as well as local sources contribute to acid deposition—in conjunction with the cost of controls makes agreement on equitable and efficient reduction strategies difficult to achieve. Coming to grips with these difficulties means understanding not only the scientific and technological dimensions of the acid rain controversy, but also the economic aspects (see Mandelbaum 1985).

ESTIMATING ENVIRONMENTAL BENEFITS

Formal economic analysis appears to have potential for guiding policy choices regarding the acid rain problem, but it has significant limits as well. Crocker and Regens (1985) note that a few attempts have been made to assess the benefits and costs of acid rain mitigation,[1] but efforts in this direction give us reason for pause. For example, Crocker, Tschirhart, and Adams (1980) estimated the maximum annual economic benefits of eliminating all acid deposition effects on current economic activities in the eastern third of the United States. As table 12 reveals, losses in the billions of dollars could be prevented. Yet, these estimates are based primarily on effects whose magnitude is still undetermined.

Unfortunately, vague statements about potential threats of environmental damage and limited quantification of biological effects are of little use when attempting to estimate the potential benefits of acid rain controls as a basis for regulatory decision making. As Kneese notes (1984, 113), "This number cannot be taken very seriously, because even if it were correct, in all other respects it neglects the large adjustments in demand and supply which would accompany the types of charges contemplated." Moreover, while some biophysical and chemical data are available, there are significant gaps in our knowledge of dose-response functions for the array of potential direct and indirect effects attributed to acid rain (see table 13). In addition, questions remain about the difficulty of capturing all benefit values for acid rain damages prevented or mitigated as well

TABLE 12
Annual Maximum Economic Losses Attributable to Acid Deposition in the Eastern Third of the United States

Effects on:	Maximum Losses ($ billion)
Buildings and structures	2.00
Forest ecosystems	1.75
Agricultural crops	1.00
Aquatic ecosystems	0.25
Other[a]	0.10

Source: Crocker, Tschirhart, and Adams (1980).
Note: Estimates in constant 1978 dollars are for the potential total benefits due to the complete elimination of acid deposition effects.

a. This category includes other potential effects on human health, water supply systems, etc.

TABLE 13
Scientific Data Needed for Estimating Economic Benefits of Controlling Acid Deposition

Effects	Inventory	Dose Response
Aquatic Resources		
Sports Fishery	water availability by susceptibility, geographical area, and species type	change in fish population with varying deposition levels
Commercial Fishery	water availability by susceptibility, geographical area, and special type	change in fish population with varying deposition levels
Ecosystem	species diversity, numbers	changes in species diversity and numbers
Terrestrial Resources		
Agricultural Crops	crop pattern by geographical area	change in marketable yield with varying deposition levels
Forests	cover type, age, stocking, and size by geographical area	change in marketable yield with varying deposition levels
Ecosystem	species diversity, numbers	changes in species diversity and numbers
Buildings and Structures		
Materials	geographical distribution by type of material and by use	deterioration rate as a function of total sulfur
Historic Monuments	geographical distribution by type of material	deterioration rate as a function of total sulfur
Water Systems	geographical distribution of systems on susceptible water bodies	change in lake intake pH
Health		
Morbidity	population	sickness per $\mu g/m^3$ $SO_4^=$
Mortality	population	deaths per $\mu g/m^3$ $SO_4^=$
Visibility	population	change in km of visibility per $\mu g/m^3$ $SO_4^=$

Source: U.S.-Canada Work Group 1 (1982).

as which damages may be irreversible. Another limiting factor is that the rate of damage may not be linearly related to deposition inputs.

Such analyses, however, can help to identify where the greatest potential benefits of acid rain control strategies are likely to be (see Regens and Donnan 1986; Crocker and Regens 1985). As a rank-ordering device, current economic analyses also can serve to prioritize categories of effects for additional research aimed at establishing dose-response functions as a basis for more sophisticated damage estimates. From an economic perspective, existing findings suggest that emphasis should be placed on defining impacts of acid rain on materials and on forest ecosystems. Effects on aquatic ecosystems, while triggering initial scientific inquiry and public concern, appear to merit less emphasis from a strictly economic standpoint.

In fact, while Crocker and his colleagues are confident that the rank-ordering of benefit categories is accurate, their $5 billion damage estimate is viewed properly as illustrating a methodology for calculating such a figure rather than as an absolute value[2] (see U.S.-Canada Work Group 1 1982). Crocker (1982, 24) makes this point when he notes that the "scant knowledge" relating to the physical and biological changes induced by exposure to acid deposition is matched only by the poor knowledge about the "price, activity, and location responses of economic agents to these system changes." He concludes that "any estimate right now of the total benefits of controlling acid deposition appears foolhardy." Those estimates of benefits that currently are available, therefore, are most appropriately understood as indicating what is most important to the various parties to the ongoing policy debate. They also may serve to highlight potential tradeoffs.

Estimating Control Costs

Cost estimates for any reduction strategy also depend upon a variety of assumptions. It is necessary to speculate about the effectiveness of various control options, future trends in emissions, the political feasibility of the different options, the stringency of current and future environmental regulations, levels of economic activity, energy prices, and technological innovation. Clearly, uncertainty surrounds each factor. Nonetheless, information about control costs is somewhat more tangible, at least in an aggregate sense.

Technological control measures are available, as we have seen, for reducing emissions of the major causes of acid deposition. Technology exists to reduce manmade emissions of SO_2, NO_x, and volatile organic compounds (VOCs). If one assumes no change in existing environmental regulations, the current level and geographical distribution of SO_2 emissions by source category, such as utilities, industrial boilers, and smelters, should be roughly stable through the year 2000. Future trends for emissions of VOCs and NO_x are less certain. Total VOC emissions will likely decline until 1990, primarily due to significant reductions in emissions from motor vehicles. After 1990, conventional wisdom holds that VOC emissions will then perhaps approach 1980 levels by the year 2000 because of increased industrial activity. The rate of growth in NO_x emissions should continue to decrease until 1990. Total NO_x emissions in the year 2000, however, are projected to exceed 1980 levels. It is important to be cautious in estimating future NO_x emissions, which are extremely dependent on assumptions about vehicle miles traveled and emissions from industrial sources. As a result, future estimates should be viewed as constituting "best judgment" projections rather than absolute approximations (U.S.-Canada Work Group 3B 1982).

Over the long term, perhaps forty or more years, there could be a major reduction in SO_2 emissions as existing sources are replaced by facilities subject to new source performance standards. The NSPS program was and still is envisioned by EPA to be a long-term strategy for limiting total emissions of pollutants into the atmosphere. As capital stock turns over, NSPS ensures that emissions will not increase disproportionately with industrial growth. It also ensures that individual states will not compete for new industry by enacting standards different from those of their neighbors. Making the standard apply only to new sources also protects existing markets for coal. Thus, it avoids the employment dislocations that would result from a wholesale shift to low-sulfur coal by existing sources constructed before 1971. However, as Regens notes (1983, 109):

Because NSPS involves determining standards for individual industry categories, it is not the most economically efficient way to limit or decrease total emissions. The strategy also ignores the availability of cost-effective emission reductions from existing sources. Moreover, although the philosophy behind stringent performance standards for new sources is sound, the regulations have had less impact on *total* emissions than was originally expected. Changes in energy demand resulting from oil price increases have altered the economics of replacing those "old" power plants that are currently among the largest sources of pollution. Thus, the rate of plant turnovers has consistently failed to meet the projections used in promulgating the standards. . . .

Some critics also maintain that the most recent NSPS may be responsible for exacerbating the existing economic disparity between existing and new sources, thereby decreasing further the rate of plant turnovers. Although EPA analyses indicate that the decrease in turnover attrributable to the latest NSPS is small relative to more overt economic parameters such as high interest rates and the generally poor financial condition of the utility industry, a programme regulating new sources stringently and existing sources leniently is more applicable to a period of strong economic growth when some economic efficiency can be sacrificed for administrative and enforcement simplicity. The strategy is also based on imposing costs in the future rather than in the present. Imposing significantly different standards between existing and new sources also promotes more subtle disparities. For example, the scrubbing requirement for new power plants imposes variable costs much higher than those for existing plants. . . . As a result, older dirtier plants are operating at a higher capacity to minimize total generating costs. In effect, the NSPS control system is providing less of an emissions reduction than society should be getting for the money which it is investing in air pollution abatement.

The above observations call into question the argument that the mere passage of time will solve the acid rain problem. In fact, it remains debatable whether replacing existing sources with new ones between 1990 and 2030 will result in emissions levels sufficient to reduce acid deposition loadings to environmentally acceptable targets. Such an expectation rests on three key assumptions. First, growth rates in the electric utility and other major emitting sectors would have to remain relatively low in comparison to historical rates. Second, technological advances would have to permit the adoption and implementation of more stringent NSPS, or innovative incentives for emissions control would have to be developed, thereby reducing aggregate emissions. Finally, no significant adverse ecological effects, especially irreversible damage, could occur within the next thirty to forty years. If these assumptions were to prove valid, then the opportunity costs of achieving additional reductions far exceed the monetary value of known, as opposed to potential damages (Regens 1984). This may seem obvious, but understanding that regulatory decisions leave us with fewer resources to use elsewhere is an important insight. Deciding about acid rain controls forces political leaders to confront a set of hard choices in which the more they allocate to acid rain controls, the fewer resources are available for pursuing other policy goals.

Given the uncertainty regarding dose-response functions, it is also possible to argue that unless large reductions in emissions, especially sulfur dioxide, occur in the next ten to fifteen years, widespread but not necessarily irreversible damage may occur. Proposals for imposing control strategies now focus on reducing SO_2 emissions to achieve reductions in sulfate deposition. This is the case because of the greater difficulty of realizing significant NO_x emissions reductions and the uncertainty over whether nitrate activity is as harmful as sulfate acidity (see Fay et al. 1983).

Because of economies of scale for pollution control efforts in the utility sector vis-à-vis the industrial sector involving fuel purchase, control technology, and transportation, it appears that requiring SO_2 reductions from utilities instead of industrial sources is relatively more cost-effective. Electric utility sources produced an estimated 65 percent of total SO_2 emissions in 1980 for the United States and are projected to dominate future trends (U.S.-Canada Work Group 3B 1982). Coal-fired power plants are the primary source of those emissions. But, because most plants are located in attainment areas, compliance with the current sulfur dioxide national ambient air quality standard is not likely to reduce significantly SO_2 emissions from existing sources.

As a result, it is important to consider the costs of a possible SO_2 reduction program. Table 14 indicates that it is generally more cost-effective to switch to lower sulfur coals or residual oil rather than to employ flue gas desulfurization. However, such switching poses potential social and economic problems: many coal miners stand to lose their jobs due to coal market shifts. Because the sulfur content of coal varies among U.S. coal-producing regions (see figure 12), "any emission reduction strat-

TABLE 14
Incremental Costs of Strategies for Reducing SO_2 Emissions

	$ per Ton	
Coal Cleaning		
Northern Appalachian and eastern Midwest coal		$50–600
Southern Appalachian coal		$700–1,000
Utility Strategies[a]		
Fuel Switching		
From high- to low-sulfur coal		$250–350
From high- to medium-sulfur coal		$350–400
From medium- to low-sulfur coal		$400–500
From high- to low-sulfur residual oil		$300–400
Flue Gas Desulfurization (FGD)		
From unscrubbed high to scrubbed high-sulfur coal		$400–600
From unscrubbed medium to scrubbed medium-sulfur coal		$600–1,500
From unscrubbed low to scrubbed low-sulfur coal		$1,800–3,000
Limestone Injection Multistaged Burners (LIMB)[b]		
High-sulfur coal	$250–500	$200–350
Medium-sulfur coal	$250–1,100	$250–700
Low-sulfur coal	$600–2,000	$500–1,200
Industrial Strategies[c]		
Fuel Switching		
From high- to low-sulfur coal		$250–350
From high- to medium-sulfur coal		$350–400
From medium- to low-sulfur coal		$400–500
From high- to low-sulfur-residual oil		$300–400
Flue Gas Desulfurization (FGD)		
From unscrubbed high to scrubbed high-sulfur coal		$400–600
From unscrubbed medium to scrubbed medium-sulfur coal		$600–1,500
From unscrubbed low to scrubbed low-sulfur coal		$1,800–3,000

Source: U.S. Environmental Protection Agency (1983).

a. Representative costs for 500 MW power plant. Costs will vary for each region and year.

b. Removal of SO_2 for retrofits expected to be between 50 and 60 percent.

c. Representative costs for 170 MM Btu/hr industrial boiler. Costs will vary for each region and year.

egy that relies even in part on 'switching' to lower sulfur coals can potentially affect the regional distribution of coal production and employment" (U.S. Office of Technology Assessment 1984, 190). Such a strategy is likely to result in increased employment and economic activity in low-sulfur coal areas and losses in high-sulfur areas. Estimated job losses in 1990 from actual 1979 levels in high-sulfur areas, assuming 50 to 75 percent fuel-switching in order to achieve a 10 million ton reduc-

Figure 12
Quantities and Sulfur Content of Coals Produced for Electric Utilities by State, 1980

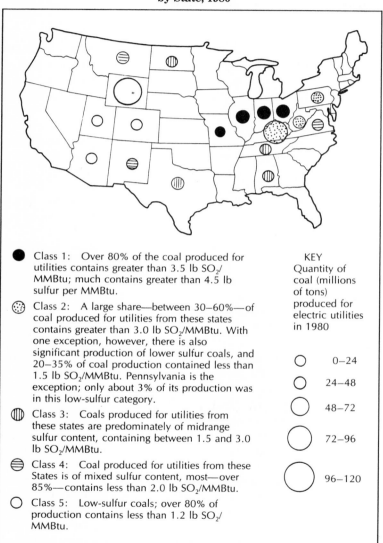

- Class 1: Over 80% of the coal produced for utilities contains greater than 3.5 lb SO$_2$/MMBtu; much contains greater than 4.5 lb sulfur per MMBtu.
- Class 2: A large share—between 30–60%—of coal produced for utilities from these states contains greater than 3.0 lb SO$_2$/MMBtu. With one exception, however, there is also significant production of lower sulfur coals, and 20–35% of coal production contained less than 1.5 lb SO$_2$/MMBtu. Pennsylvania is the exception; only about 3% of its production was in this low-sulfur category.
- Class 3: Coals produced for utilities from these states are predominately of midrange sulfur content, containing between 1.5 and 3.0 lb SO$_2$/MMBtu.
- Class 4: Coal produced for utilities from these States is of mixed sulfur content, most—over 85%—contains less than 2.0 lb SO$_2$/MMBtu.
- Class 5: Low-sulfur coals; over 80% of production contains less than 1.2 lb SO$_2$/MMBtu.

KEY
Quantity of coal (millions of tons) produced for electric utilities in 1980

- 0–24
- 24–48
- 48–72
- 72–96
- 96–120

Source: U.S. Office of Technology Assessment (1984).

tion, range from 23,000 to 60,000, producing total economic losses from $1.6 to $4.6 billion in 1981 dollars (see U.S. Office of Technology Assessment 1984; UMW 1983). The indirect economic impact of acid rain controls may also extend to industries whose industrial processes rely heavily on electricity. Recent studies suggest that utility compliance strategies can have substantial impacts on the costs of electricity-intensive sectors of the economy such as steel and textiles (Temple, Barker, and Sloane, Inc. 1985; Arthur D. Little and Energy Ventures Analysis 1984).

While the engineering economics of limestone injection multistage burners seem promising, LIMB commercialization does not appear likely before the mid-1990s. In addition, because of the time required for a new technology to move from the pilot to demonstration to commercial stage, LIMB is not going to be a major option for retrofitting existing power plants before then, even if its development and demonstration expenses are substantially underwritten by the federal government.

To estimate the actual costs of an SO_2 emissions reduction program in terms of control costs, coal market shifts, or electricity rate increases, analysts must specify a number of prior conditions. Both the size of the emissions rollback in terms of tons per year and the geographical area, such as thirty-one or forty-eight states in which those emissions reductions are required, must be defined. The schedule for the implementation of additional controls is also a major determinant of program costs. The imposition of further controls now, for example, requires the use of currently available FGD technology. To delay actual reductions until after the early 1990s allows consideration of possible new technologies such as LIMB as a potential option. The presumed advantages of such a delay, however, must be weighed against the potential environmental damage in the interim. Additional parameters—such as the permissibility of emissions trading, requirements for NO_x caps, NO_x substituting for SO_2 reductions, and protection of coal miners by restricting or prohibiting fuel switching—also affect cost estimates. Finally, an actual strategy, if implemented, might well include mitigation measures as well as emissions reductions, thereby influencing the ultimate total cost.

A number of studies provide information about the overall

costs to representative emitters for reducing the precursors of acid deposition (Rubin 1983; U.S. Office of Technology Assessment 1982; U.S.-Canada Work Group 3B 1982; McGlamery and Torstick 1976). Those analyses provide substantial insight, if not definitive answers, into the costs under alternative control regimes, especially for SO_2 reductions. None of the cost studies explicitly considers the unique problems of the utility industry, how these problems influence its institutionalized habits and modes of thinking, and therefore the observed costs on which the aforementioned studies are founded. Under some circumstances, control costs in the industry are likely to be above those that would be experienced by profit-maximizing, perfectly competitive producers of identical outputs (Goldberg 1976; Averch and Johnson 1962). The aggregate as well as utility-specific size of the increase, if any, is unknown.

Nonetheless, conclusions about the costs of achieving aggregate SO_2 emissions reductions in the utility sector have been quite uniform. Annualized cost estimates in 1982 dollars range from $1 to $2 billion for a 40 percent reduction and from $2 to $4 billion for a 50 percent reduction. Cost consequences of $5 to $6 billion are estimated for 66 to 75 percent reductions. The estimated average increases in electricity rates to accomplish such a reduction in utility industry SO_2 emissions have ranged from 1.4 percent for a 4 million ton rollback to 8.0 percent for a 12 million ton rollback. Naturally, rate increases as well as control costs for individual utility systems may be substantially greater than the average values.

A broad consensus exists that using economic incentive systems such as marketable emissions permits to control air pollution would be as much as an order-of-magnitude less costly than the current SIP-based system (Seskin et al. 1983; Atkinson and Tietenberg 1982). Recent studies of acid rain controls, however, suggest that the cost consequences of alternative strategies for SO_2 reductions for acid deposition purposes have only minor differences. State Implementation Plan systems tend to be as effective as economic incentive systems in addressing long-range transport (Atkinson 1984; U.S. Office of Technology Assessment 1982; Silverman 1982).

Atkinson (1984, 22) attributes this result mainly to the greater aggregate quantity of emissions that localized economic incentive strategies allow to meet a given local ambient standard. Emissions are distributed spatially such that the dispersal properties of the local atmosphere are used more effectively. These greater emissions provide more material for long-range transport. Given the gradual way in which SO_2 combines with other atmospheric constituents to form $SO_4^=$, each point source makes an incremental contribution to a regional sulfate problem, instead of imposing the primary sulfate burden on the surrounding area. Moreover, each source's impact is inversely related to its impact on the local ambient problem. The more SO_2 that returns to earth close to the source, the less remaining for chemical transformation to $SO_4^=$ over longer periods and across larger areas. When the material is removed to meet ambient standards that account for long-range transport, the cost advantage of the economic incentive strategies is drastically reduced.

While emissions reduction strategies focus on controlling air pollution at its source, mitigating the adverse effects of acid rain at receptor sites is an alternative or complementary policy option. For example, liming has been suggested as one mitigation strategy for inhibiting the aquatic effects of acidification. Extensive liming has been done in Sweden. Since 1977, about 3,500 bodies of water have been treated, with approximately 50 percent success from a chemical point of view.[3] To date, however, only small-scale attempts at restoration using liming techniques have been made in the United States. Costs for individual lakes are likely to vary depending on the cost for materials and application, lake size, and preapplication planning and postapplication monitoring (see Fraser et al. 1984). Thus, although the efficacy of mitigation strategies such as liming and/or accelerated research has been debated, congressional attention has been focused on the merits of emissions reductions strategies based on cost or equity criteria.

Just as there are various plausible interpretations of the science of acid rain, there are also contending judgments about the economics. The two alternatives are as follows. The rationale for deferring additional action argues that the benefits of control are largely intangible, while the costs are substantial. On the other side, those who support taking immediate action point to potential damage to the environment and reduced visi-

The Economic Dimension 97

bility that would result from delay—effects that would undoubtedly have economic significance. This debate illustrates how existing information can be used to construct a policy rationale for deferring action and maintaining the status quo, or taking further action now to initiate controls on acid deposition. Let us consider how each of these alternatives might be financed.

Alternatives for Financing Control Costs

Acid rain forces policymakers to confront a typical environmental controversy. Popular perceptions of the degree of environmental risk may be more powerful motivations for government intervention than actual scientific evidence. Moreover, the United States has a long tradition of undertaking regulatory action to demonstrate public concern or simply to do something in the face of crises. And, as a society, we often implement risk-reduction strategies on the basis of only fragmentary evidence of hazards themselves or of the relative costs and benefits of other strategies (see Crandall and Lave 1981). As a result, the financing issue, including questions of equity, has increasingly become the focus of both the public policy debate and the symbolic politics surrounding the allocation of costs for an acid rain control program.

The Electric Utility Rate-making Process

In one of the more commonly discussed financial alternatives, Congress or EPA would set emission targets. Each state would then allocate its share of the required reductions among the electric utilities in the state. Two scenarios exist for allowing the individual utility systems to determine how to achieve their reduction quotas. Under one scenario, each utility would decide on an appropriate strategy to achieve the necessary emissions reduction. The utility would then ask its state public utility commission (PUC) to permit rate increases necessary to offset the additional control costs. This scenario leaves the actual choice of specific compliance approach, such as scrubbing via FGD retrofit, fuel switching, or physical coal cleaning, to the discretion of the utility and/or the various state governments. The second scenario also would use the rate-making process to generate capital for financing acid rain controls. But the actual array of control options to be employed in order to accomplish the SO_2 emissions reduction would be statutorily mandated.

Under either scenario, the affected utilities generally would have to pay for construction costs prior to recovering all or a portion of those costs via their rate structures. The delay before a utility is reimbursed for a pollution control expenditure varies considerably by state. For capital expenditures, the length of the delay depends primarily on whether or not utilities are permitted to charge ratepayers for financial costs associated with current construction. Although incorporating the costs for construction-work-in-progress (CWIP) into electricity rates is generally not allowed by most PUCs while new plants are being built, it is frequently permitted for the costs of installing pollution control equipment. These costs include such hardware as FGD retrofit or upgrading an electrostatic precipitator to handle low-sulfur coal.

Electricity consumers, on the other hand, would not begin paying for emissions reductions until shortly after the control technology to net those reductions became operational, unless CWIP is allowed. The capital costs for acid rain controls typically would be amortized over a fifteen-year period, although the amortization period could range anywhere from ten to thirty years. Operating costs for pollution abatement would be recovered on an annual basis. If a utility were not fully reimbursed for its acid rain control investments, the remaining cost increment would be shifted to its owners and/or investors.

Because a program for reducing SO_2 emissions diverts capital from availability for investment in other sectors of the economy, the electric utility rate-making process is one of the more economically efficient mechanisms available for funding such a program. An efficiently implemented 8 million ton regional rollback in SO_2 emissions, such as S. 3041 proposed, is estimated to cost $3.7 billion, on an annualized basis, in 1995.[4] Utilities would pay $3.1 billion of this total on a yearly basis and other industries would pay an additional $.6 billion per year (see ICF, Inc. 1983). The incremental administrative costs should be relatively low, especially compared to other approaches, since the rate-making process relies on an established system. In addition, reliance on the rate-making process results in imposing control costs on those utilities that are required to reduce emissions according to the polluter-pays principle underlying most existing U.S. environmental statutes. According to microeconomic theory, electricity generators (utilities) and consumers (ratepayers) would receive information about the true costs—including the environmental impacts—associated with their production and consumption of electricity. Presumably, this would cause them to adjust their behavior accordingly.

Such an approach, however, is not without disadvantages. Reliance on the utility rate-making process to finance acid deposition controls would impose the bulk of the costs of SO_2 emissions reductions on utilities burning high-sulfur coal. Table 15 indicates that electricity consumers in some states would receive relatively large rate increases in the initial years of such a control program. For example, a 10 million ton reduction might increase utility rates in the Midwest by an average of 7.4 percent. While those midwestern states could partially or totally protect their low-income consumers from absorbing such an increase with some form of assistance, such a financial subsidy would require allocating a greater share of the overall control costs to more affluent individuals and/or industrial customers.

A rate increase of that magnitude could harm the Midwest's already depressed economy as well as individual ratepayers in the region. For example, many long-term investments have been made based on past electricity rates. Those utilities that would pay the most for controls often serve regions whose major industries were hard hit by the recent recession. If their financial condition does not improve between now and the mid-1990s, substantial electricity rate hikes could further hasten their decline relative to the rest of the country. In addition, if rate increases are substantial, which is as much a relative as an absolute judgment, utility systems may seek to reduce their compliance costs by fuel switching, importing electric power from Canada, or reducing generation instead of scrubbing to achieve required emissions reductions.

To the degree that utilities are able to reduce their SO_2 emissions by switching from reliance on high-sulfur local coals, coal production in the Midwest is likely to decline for several

TABLE 15
Electricity Rate Increases for a 10 Million Ton Reduction in SO_2 Emissions
(in percent)

Region	State	%
New England	Connecticut	6.1
	Massachusetts	6.1
	Maine	3.2
	New Hampshire	3.2
	Rhode Island	6.1
	Vermont	3.2
X̄		4.7
Mid-Atlantic[a]	Delaware	3.2
	Maryland	3.2
	New Jersey	2.3
	New York	2.8
	Pennsylvania	6.5
X̄		3.6
Midwest	Iowa	3.3
	Illinois	0.7
	Indiana	13.5
	Michigan	3.6
	Minnesota	−0.9
	Missouri	13.0
	Ohio	17.8
	Wisconsin	8.3
X̄		7.4
South	Alabama	−2.1
	Arkansas	2.9
	Florida	3.7
	Georgia	4.6
	Kentucky	10.8
	Louisiana	0.4
	Mississippi	12.8
	North Carolina	1.2
	South Carolina	1.2
	Tennessee	10.1
	Virginia	2.6
	West Virginia	6.0
X̄		4.5

Source: Adapted from Wetstone (1983).
Note: Table reflects costs based on percent change in electricity rates in 1990 for a first year revenue requirement on a composite bill assuming intrastate trading.
a. Excludes the District of Columbia.

reasons. The current electric rate-making system encourages minimizing capital expenditures instead of generation costs per kilowatt hour (kwh). If utilities fail to receive a rate of return commensurate with the risks involved in capital expenditures, options with relatively high operating and maintenance (O&M) costs are preferred. For example, some electric utilities have been reluctant to convert their oil-burning units to coal, even though the long-term fuel saving for such a power plant conversion would more than offset the increased capital costs at the front end (Brenner 1983). Similar considerations would cause a utility to stop using high-sulfur coal rather than install a FGD system. Moreover, given widespread skepticism about FGD among utility executives, this is likely even if the costs per kwh for scrubbing were cheaper.

Finally, because of the rate-making system, control costs are heaviest in the early years of the payback period. This feature could make the first year's rates as much as 50 percent higher than the long-term average costs of the program. Rate-making reforms, such as agreements by PUCs to allow CWIP, could equalize rate increases over time. However, because the normal rate-making process forces utilities to raise the funds for financing acid rain controls through the capital market, utility systems with low allowable rates of return, poor growth prospects and/or heavy debts will probably be unable to finance controls without some public-sector assistance. As a result, irrespective of the science of acid rain, those utility systems are likely to fight control proposals based on economic reasons alone.

The political attractiveness of an acid rain trust fund or some other alternative to the normal electric utility rate-making process is illustrated in this passage from *Fortune* magazine:

For now, the stock market seems untroubled by the specter of acid rain legislation. "There are psychological issues that haven't moved the stocks yet," says Margery Obrentz, a security analyst at Kidder Peabody & Co., "but as Congress gets more serious about this, we expect the stocks to move a little."

Perhaps the market is so untroubled because no one really expects Congress to slap an unproductive $10-billion or $20-billion capital expense on the utility industry. Not that acid rain legislation won't pass; it probably will. But sentiment is strong against saddling one industry or region with the whole cost, especially since smelters, steel mills, auto plants, and home furnaces all contribute to the acid brew. So it may well be that one of several spread-the-cost schemes under consideration will shift the burden at least partially off utility shoulders (Magnet 1983, 60).

Alternatives to the Electric Utility Rate-making Process

In general, alternatives to the normal process of setting utility rates are less economically efficient. However, those options may be preferred for several reasons. First, most allocate the costs of financing an SO_2 emissions reduction program over a broader segment of the population. Therefore, no one group would receive substantial electricity rate hikes. An argument can be made that it is less disruptive and, therefore, more socially desirable to require a number of small adjustments rather than a few large ones.

The alternatives involve raising all or part of the funds for financing controls on the basis of electricity production, emissions, fuel use, general revenues, or a combination of these sources. All of these options would give two more congressional committees, the House Ways and Means Committee and the Senate Finance Committee, jurisdiction over legislation for acid rain control. This might further complicate consensus-building around a politically feasible control program. On the other hand, the options also provide opportunities to target funds on the basis of additional social or economic goals. For example, revenues could be used to provide subsidies to severely affected ratepayers, especially low-income consumers. They also might be used to mitigate the hardship caused by job losses in the Midwest due to reduced high-sulfur coal production.

Generation Fee Proposals for a generation fee on electricity production continue a recent trend in environmental legislation. Both the Superfund for hazardous wastes and the nuclear waste storage programs rely on forms of generation fees to subsidize cleanup efforts (see Wetstone 1983). The most likely scenario for implementing a generation fee requires utilities to collect the revenues as a kilowatt hour surcharge. The fee could be imposed on all electricity generation or only on fossil fuel electricity production (conventional steam). Applying it to the latter seems more logical since hydro and nuclear capacity are not sources of SO_2 emissions while fossil fuel–fired power plants are. The money collected by the fee would be invested in a trust fund administered by either the Treasury Department or EPA. Income from the trust might be used to fund capital costs, O&M costs, some combination of the two, or other targeted purposes such as coal miner or ratepayer assistance.

The fees proposed to date range from 1 to 3 mills per kwh.[5] A typical residential customer uses approximately 750 kwh of electricity per month.[6] A three mill/kwh fee would increase a monthly bill by about $2.25, although some slight regional differences would exist in the actual percentage increase reflected in utility rates by adopting a generation fee. If tax revenues did not fully subsidize control costs by covering only capital but not operating and maintenance expenses, electricity users would bear additional rate increases beyond the generation fee. Consumers would also be subject to an indirect cost—higher prices for goods produced by those industries subject to the tax. However, there are relatively few products, other than possibly steel and textiles, for which the cost of electricity is a major component of their total cost.

In spite of the low rates, a generation fee, with the rate set in either real or nominal terms, can produce massive revenues.[7] For example, using projected 1985 fossil fuel–based electricity generation, a one mill/kwh fee would raise approximately $1.2 billion per year in the thirty-one state region encompassing the eastern United States. If the fee were imposed across forty-eight states, the estimated revenues would increase to $1.75 billion.[8] Clearly, a generation fee is an effective means for spreading control costs over a larger base, making the impact on individual regions or states relatively small. As a result, the approach should not significantly affect competition between electricity and other kinds of energy. A generation fee is also relatively simple to collect. It is essentially equivalent to a sales tax on electricity consumption, so it is easily implemented.

As is true for any of the options, the generation fee concept does have some problems. Although the generation fee is based on output rather than input, it still would make low-sulfur fuels more attractive than high-sulfur coal. As noted above, utility companies would still prefer to switch fuels instead of paying for the O&M costs of a scrubber. As a consequence, a premium would exist for low-sulfur fuels unless they were proscribed by control legislation. The most serious weakness of the generation fee approach is its failure to give credit for current and past pollution abatement efforts. Table 16 demonstrates this problem. Since an individual state's liability becomes a function of how much electricity it produces, states with high generation and relatively clean units would not be treated differently from high electricity-producing states with limited prior controls. As a result, those consumers who have already paid for FGD systems or low-sulfur fuels would pay again to subsidize ratepayers of utilities with relatively uncontrolled units. The criticism remains valid even if the revenues are to subsidize control costs rather than fully funding them.

While a graduated rate based on emission rates provides some incentive for electric utilities to operate their less polluting units more intensively and does offer some recognition for past actions, it would create a somewhat artificial, although probably marginal, economic discontinuity in decisions about load management. Both nuclear power plants and hydroelectric facilities can be substituted only for fossil fuel–fired plants on the margin in most utility systems. Nuclear plants are baseload capacity units and hydroelectric capacity is constrained by the amount of water available for electricity generation. As a consequence, any differential in load management of fossil fuel power plants is likely to stem from factors other than a generation fee surcharge.

Btu Tax A Btu tax is similar to a generation fee in many respects.[9] But using energy consumption as the basis for allocating how revenue is generated makes it possible to reach an even broader base. This follows because a Btu tax could be applied to the use of all fossil fuels or to consumption above a given level. For example, industrial boilers, process heat applications, and motor vehicles, as well as utility boilers, consume fossil fuel in the form of coal, natural gas, residual oil, distillate, or gasoline. That fuel use could be taxed on a Btu basis. As a result, if industrial and utility tax rates were the same, utilities would provide only approximately two-thirds as much revenue, reflecting their share of total U.S. fossil fuel consumption. For example, a tax of 6.13 cents/10^6 Btu (real) would be roughly equivalent to a 1 mill/kwh (real) generating fee. The key difference is that about 38 percent of the revenues raised by a Btu

TABLE 16
Distributional Impact of Selected Options for Financing Reductions in Acid Deposition
(in percent)

	Generation Fee[a]	Btu Tax[b]	Emissions Fee[c]
Alabama	2.56	2.02	2.86
Alaska	0.13	0.41	0.07
Arizona	1.92	1.06	3.39
Arkansas	0.85	0.92	0.38
California	5.51	7.98	1.68
Colorado	1.32	1.24	0.50
Connecticut	0.62	0.88	0.27
Delaware	0.47	0.34	0.41
District of Columbia	0.02	0.11	0.06
Florida	4.78	3.30	4.12
Georgia	2.98	2.27	3.16
Hawaii	0.35	0.37	0.22
Idaho	—[d]	0.27	0.18
Illinois	3.93	4.96	5.54
Indiana	3.82	3.46	7.56
Iowa	1.10	1.30	1.24
Kansas	1.36	1.52	0.84
Kentucky	3.27	1.91	4.22
Louisiana	2.51	4.51	1.14
Maine	0.12	0.32	0.36
Maryland	0.97	1.36	1.27
Massachusetts	1.60	1.63	1.30
Michigan	3.25	3.64	3.41
Minnesota	1.16	1.54	0.98
Mississippi	0.84	1.04	1.07
Missouri	2.75	2.07	4.90
Montana	0.30	0.43	0.62
Nebraska	0.50	0.68	0.28
Nevada	0.80	0.44	0.91
New Hampshire	0.25	0.26	0.35
New Jersey	1.15	2.45	1.05
New Mexico	1.30	0.92	1.01
New York	3.56	4.71	3.56
North Carolina	3.57	2.16	2.27
North Dakota	0.78	0.48	0.40
Ohio	6.10	5.53	9.97
Oklahoma	2.44	1.85	0.45
Oregon	0.10	0.66	0.23
Pennsylvania	5.78	5.53	7.61
Rhode Island	0.05	0.20	0.06
South Carolina	1.29	1.10	1.23
South Dakota	0.15	0.26	0.15
Tennessee	2.75	1.91	4.05
Texas	11.71	12.10	4.81

	Generation Fee[a]	Btu Tax[b]	Emissions Fee[c]
Utah	0.63	0.71	0.27
Vermont	—[d]	0.10	0.02
Virginia	1.10	1.67	1.36
Washington	0.40	1.16	1.02
West Virginia	4.16	1.74	4.10
Wisconsin	1.43	1.71	2.40
Wyoming	1.45	0.78	0.69

Source: Computed by the authors.

a. Based on 1981 conventional steam generation (i.e., nonhydro, nonnuclear) of electricity expressed as 10^6 kwh. See Edison Electric Institute (1983).

b. Based on 1980 consumption of fossil fuels (coal, natural gas, and petroleum products) expressed as 10^{12} Btu. See U.S. Department of Energy (1983).

c. Based on 1980 SO_2 emissions estimates expressed as 10^6 tonnes. See U.S.-Canada Work Group 3B (1982).

d. State shares < .001 percent of total U.S. revenue requirement.

tax would be provided by the industrial sector. The tax would be calculated either from fuel consumption or from a surcharge on fuel purchases. From an administrative standpoint, it would be far easier to impose a surcharge on fuel purchases than to estimate consumption. This is important in considering market-based approaches since the very costs of administering a revenue scheme might well exceed the gains from eliminating economic inefficiency (see Coase 1960).

A Btu tax would impose a larger percentage price increase on coal than on other fossil fuels. As a result, if policymakers want to encourage coal use, relying on that option to finance acid rain controls has a major drawback. But the percentage price increases for coal, oil, and natural gas are roughly proportional to the relative adverse environmental effects that each produces. Thus, to the extent an incentive is created, it would be for the more efficient use of energy. The fewer Btu's consumed, the smaller the relative share of the tax borne by fuel suppliers and consumers (see Table 16). While a Btu tax encourages energy conservation, it further penalizes sources that have incurred energy penalties by installing such equipment as FGD systems to control emissions. However, because it potentially uses fossil fuel consumption as a base, it does provide a mechanism for reducing NO_x as well as SO_2 emissions.

Emissions Fee At least since the debate surrounding the

adoption of the 1970 Clean Air Act, economists have argued about whether an emissions fee would provide a cost-effective incentive for dealing with air pollution problems (see Freeman 1979; Mills 1978). Unlike generation fees or Btu taxes that impose equal costs on all, an emissions fee imposes a differential burden. In one sense, it embodies the "polluter-pays" principle. Pollution abatement costs are allocated to sources according to their prior success in controlling emissions. As a result, electricity prices better reflect the true cost—including the environmental impact—of production. Presumably, an emissions fee approach to financing acid rain controls would be based on the number of tons of SO_2 produced by specified sources. Emissions could be either monitored or calculated from data on fuel inputs and emission coefficients. Each represents a substantial administrative undertaking, especially were continuous emissions monitoring (CEM) to be adopted.

For example, a $120 per ton SO_2 emissions fee imposed on major sources would raise roughly as much money as a one mill/kwh generation fee. It would also induce a few sources to reduce their emissions, but it concentrates the revenue burden on those areas that will later be expected to incur the cost of controls. On the other hand, the same areas will presumably receive many of the benefits when those revenues are disbursed.

Given the barriers to implementation, this approach to financing acid rain controls remains attractive for other reasons. An SO_2 emissions fee, unlike a Btu or generation fee, would address directly the SO_2 loadings problem by changing emission levels and raising revenue. Economic theory suggests that the ideal point to set the fee would be somewhat lower than the marginal cost of control for capturing the last ton of emission reduction desired. In theory, sources that could reduce their emissions relatively cheaply would do so to avoid paying the fee. Other sources facing relatively large control costs would generally choose to pay the fee. A variation on this theme would be to establish allowable emissions in a particular area and then sell or give the right to produce a portion of that total to current polluters. They could then resell those rights if they desired (see Liroff 1986).

While attractive to economists, the emissions fee concept is not without drawbacks. Those disadvantages primarily involve its implementation and economic impact on domestic coal markets. As noted above, an SO_2 fee would be difficult to administer. Environmental authorities can estimate an individual power plant's maximum emissions rate in a general way, but determining its total emissions is difficult, if not impossible, without continuous emissions monitoring. Thus, it would be necessary to invest in either a widespread CEM effort or an emissions auditing program. The SO_2 emissions fee imposes costs in a pattern very similar, but not quite as focused, as an efficient allocation system coupled with the normal rate-making process. As a result, unless fuel switching were constrained, coal markets would be impacted.

Table 16 indicates that, under an emissions fee approach the costs of achieving SO_2 emissions reductions would remain concentrated in a few states. Some individual utility systems would still have to pass on large expenditures to their ratepayers as a consequence. Therefore the SO_2 emissions fee, like the normal rate-making process, would create great pressure for fuel switching. Moreover, it might not produce the level of geographical distribution of SO_2 emissions reductions desired. An SO_2 fee used to allocate the reduction would prevent sources from having to pay extremely high costs per ton of SO_2 removed. Instead, they could choose to pay the fee rather than make the investment in pollution abatement. As a result, the emissions fee approach provides an obvious mechanism for generating revenue, but does not necessarily force an actual emissions reduction. This problem, coupled with certain objections to the idea in principle, has led many environmentalists to oppose the emissions fee approach.

Sulfur-content-in-fuel tax A sulfur-content-in-fuel tax is essentially comparable to an SO_2 emissions fee. However, it differs in one important way. It is based on sulfur input instead of output. As a result, a sulfur-content-in-fuel tax would be imposed on the sulfur content of fossil fuel purchases rather than SO_2 emissions. Nevertheless, it probably would have to be based on potential SO_2 emissions per million Btu's in order to be standardized among fuels. Once again, unless changing fuel was prohibited, most sources probably would prefer to shift to lower sulfur fuels to avoid the tax. As with some of the other options, unless some type of credit were provided, the tax would also penalize FGD-equipped sources subject to the tax. Yet they would already have made substantial investments in air pollution control equipment that reduces emissions but carries with it an energy penalty. This is the case because FGD systems require fuel to operate, thereby increasing fuel consumption but not electricity generation. With the credit, the impact of a sulfur-content-in-fuel tax would be virtually identical to that of an SO_2 emissions fee. Without the credit, however, FGD systems are not likely to be installed for retrofitting utilities. In theory, such a credit could be set at such a level as to leave decision makers indifferent with regard to scrubbing and fuel switching, or it could function as a positive incentive to encourage scrubbing of high-sulfur coals.

Finally, it might be less complex administratively to implement a sulfur-content-in-fuel tax than an SO_2 emissions fee because sulfur inputs measured in terms of liquid and solid fossil fuels can be established in physical units more straightforwardly than can outputs measured as emissions from dispersed point and mobile sources. However, whether this would reduce administrative complexity has not been demonstrated empirically; this could be tested if acid deposition controls were financed with this particular option. As with an emissions fee, raising revenues with a sulfur-content-in-fuel tax does not guarantee that SO_2 emissions reductions will maximize environmental instead of economic benefits.

Federal Budget Outlays Each of the above alternatives, including the normal utility rate-making process, ties the mechanism for revenue generation to some aspect of energy use and its residual air pollution effects. It is possible, at least conceptually, to divorce totally or partially acid deposition control costs and the rise in revenues necessary to fund those controls. The cost of acid rain controls could be paid directly by the federal government as part of the federal budget. Approaching the financing question in this way spreads control costs over the largest possible base, thereby minimizing individual liability. It also does not require a significant expansion of existing administrative structures. The process is administratively simple, since the existing mechanism for raising federal funds would be used.

These considerations seem to make relying on direct federal budget outlays an attractive option. Turning to general revenues implies both that acid deposition control is an important national concern and that there is a consensus that ranks it as a major priority. We simply do not have enough evidence to draw such a conclusion at this time. In fact, some aspects of reliance on the federal budget to subsidize controls call into question the option's political feasibility. The use of direct federal budget outlays represents a clear departure from the "polluter-pays" principle unless the concept is construed so broadly as to lose any analytical meaning. Relying on general revenues is completely unconnected with prior abatement efforts. Unlike some of the other alternatives, which can be justified on the basis of energy policy objectives, this approach also fails to change relative fossil fuel prices, thereby promoting conservation or efficient end use management. Finally, in a period marked by substantial concern over taxes and deficits, the approach forces a clear choice between increasing one or the other. These drawbacks seem sufficient to mobilize significant political opposition if not preclude using direct federal budget outlays to finance all or some part of the costs of acid rain controls.

Implications for Policymaking

From what we know of the science and economics of acid rain, and because each of the financing alternatives can be made to generate essentially equivalent revenues, policymakers recognize that the selection of an option ultimately depends on distributional considerations. Howard and Perley (1980, 185–86) succinctly state the obvious conclusion: "Acid rain is an environmental crisis with sweeping financial implications, but its solution is political." In other words, classic equity issues—who benefits and who bears the burden of the costs and risks—are likely to decide what solution is chosen. Rhodes and Middleton (1983, 7) make this point:

As acid rain control measures are proposed and debated, an important observation must be taken into account in the formulation of regulatory policy; namely, that the current unregulated acid rain situation produces *certain benefits* to some and *uncertain costs* to others. Proposed regulation of acid-causing emissions, however, promises *certain costs* for those who presently enjoy the certain benefits but *uncertain benefits* for those who presently incur the uncertain costs.

In such circumstances, the economic efficiency of any financial strategy may be of limited importance but establishing the economic benefits of reducing acid rain is central to generating broad-based support for control measures. The very different patterns of bureaucratic politics and interest group mobilization and interaction, including how they use information from the physical and natural sciences or economics, are probably more significant in determining the ultimate outcome of the acid rain debate (see Stanfield 1986b).

5
ACID RAIN AS A POLITICAL PROBLEM

Acid rain is a litmus test for the nation, a test of how well we choose to confront and solve our environmental problems. To date, the public debate over acid rain has gone as sour as the rain itself. No other controversy in recent time has so divided the nation along regional lines, nor engendered such bitter dispute among powerful competing interests.

—*Roy Gould (1985)*

AMBIGUOUS SCIENTIFIC EVIDENCE and strong disagreement as to the economic benefits of regulation are not unusual in the fields of environmental protection or occupational health and safety. In fact, in most cases, scientific evidence is a necessary but seldom a sufficient factor in making regulatory policy. More commonly, science is used to clarify issues, to isolate judgments about the costs, risks, and benefits of various policy alternatives, and to provide a framework within which complicated regulatory decisions can be made (Katz 1984; Regens et al. 1983; Crandall and Lave 1981). Quite often science serves a valuable legitimization function for decisions made on other grounds. Science also can be manipulated by bureaucracies, and at times this manipulation can take the form of outright abuse of the scientific enterprise itself (Rushefsky 1984). Thus, science can be linked to the policy process in a number of ways.

Ashford (1984), for example, has identified three major linkages between science and policy in regulating environmental hazards. First, research that precedes or justifies decisions, including toxicological, epidemiological, or other kinds of evidence, must be aggregated and analyzed to form the basis for agency determinations as to whether a hazard exists, at what level, and with what consequences for various parties-at-interest. Second, technical analysis may be needed to determine which of the primary studies should be assigned to the highest level of scientific validity and credibility, so that disputes can be resolved within the bueaucratic setting or with client groups. Finally, policymakers can be helped by legal and political analyses that try to clear up remaining uncertainties in the scientific data base, so as to reduce political uncertainty among those who decide how costs, risks, and benefits should be allocated.

Given these complexities, it is no surprise that there are substantial intra- and interagency variations in the use of scientific and technical information. Differences in professional training, agency missions, and statutory mandates contribute to wide differences in how bureaucrats use expert analyses (Kraft 1982). As a result, it is not immediately obvious how EPA's approach may differ from that of other federal agencies, since U.S. bureaucracies seem to have always maintained substantial discretionary authority in assembling, interpreting, and applying scientific information (Crandall and Lave 1981). Moreover, the key decisions will not be the bureaucracy's alone. It is not entirely clear how bureaucratic uses of information for decision making might diverge, if at all, from legislative initiatives that also affect the connection between science and policy.

Because of the ambiguous relationship between good science and the regulatory process, and because of the limited resources for scientific-technological studies and delays in bureaucratic responses, any regulatory agency will have a difficult time considering tradeoffs between benefits and harms (Ramo 1981). Yet these tradeoffs are central to successful performance. No bureau can function unless it can to some extent control how highly technical decisions create winners and losers among its clients.

Chubb (1983, 18–57) provides perhaps the most useful conceptualization of the bureaucratic strategies for making such determinations of cost-bearers and beneficiaries. He identifies three strategies. The first is corporatist, with the bureaucracy responding to beneficiary groups exclusively. The second is pluralist: the agency makes diversified overtures to both beneficiaries and cost-bearers. The third strategy is co-optive, in which bureaucrats defer to cost-bearers alone.

Bureaucratic Politics and the Environmental Protection Agency

By drawing together a number of units scattered throughout the federal government, the Nixon administration's Reorganization Plan Number Three created EPA in 1970. As an "experiment in regulatory reform," Congress attempted to narrow EPA's discretionary authority within a relatively specific legislative mandate to save the agency from "bureaucratic sloth" (Marcus 1980, 287). This reform represented a very substantial departure from, even a rejection of, the New Deal approach to public administration. According to Ackerman and Hassler (1981, 1):

The passions of Earth Day have marked our law in deep and abiding ways. Statutes passed in the early 1970s did more than commit hundreds of billions of dollars to the cause of environmental protection in the decades ahead. They also represent part of a complex effort by which the present generation is revising the system of administrative law inherited from the New Deal. The rise of environmental consciousness in the late 1960s coincided with the decline of an older dream—the image of an independent and expert administrative agency creatively regulating a complex social problem in the public interest. When Congress reacted to Earth Day, it tried to do more than clean the water and purify the air; it also sought a new shape for the administrative process—one that would avoid the use of expertise as an excuse for inaction and would protect agencies from capture by special interests.

It is no small irony that the debate surrounding acid rain policy has featured vigorous accusations of expertise-based inaction, which reformers were so intent on eliminating from the original design of EPA.

In its early years, EPA exhibited a modified corporatist or beneficiary-oriented strategy toward most environmental issues. Several factors help explain why the new agency often deferred to its beneficiaries, the environmental groups. First, the Nixon administration, which might have been expected to try to limit the influence of environmental groups, had no special interest in environmental reform. The administration did consider, but ultimately rejected, a set of recommendations from the Ash Council that would have folded environmental protection programs into a more comprehensive Department of Natural Resources (Quarles 1976). Created as a separate entity, EPA was designed to represent environmentalists. Of course, successful agency performance depended on maintaining some balance with industrial cost-bearers, if only to acquire needed scientific and technical information and expertise. And, indeed, the agency's first administrator, William Ruckelshaus, acted as if he wished to keep some distance between himself and both of the primary parties-at-interest (Marcus 1980).

The second influence on the evolution of EPA's strategies was a set of legislative factors. As we have noted, Congress designed the agency's statutory mandate to avoid co-optation by special interests and to limit discretion. This involved shaping the entire administrative process through the establishment of clearer missions and mandates and forcing compliance through a more hard-hitting, adversarial approach (Daneke 1982). These were reforms long supported by environmental groups. In addition, Congress made it very difficult to incorporate the concerns of cost-bearers into many of EPA's procedures. The Clean Air Act, for example, explicitly forbids the use of cost-benefit analysis in setting primary standards to protect public health.

Third, the climate of the time favored the influence of beneficiaries. The late 1960s and early 1970s were characterized by substantial optimism about our society's ability to have a clean environment at relatively little cost. President Nixon himself reflected this optimism in a speech at the signing ceremony for the National Environmental Policy Act on January 1, 1970, when he said, "The 1970s absolutely must be the years when America pays its debt to the past by reclaiming the purity of its air, its waters, and our living environment" (Editorial Research Reports 1982, 139). Few anticipated either the expenses involved in pollution abatement, the pressures that expanded energy resource development and economic growth would place on the environment agenda, or the emergence of more complex environmental issues such as acid rain, toxic substances control, or significant groundwater contamination problems.

Taken together, these factors explain the agency's consistent willingness to implement a "polluter-pays" approach and the attractiveness of various technology-forcing rules to ensure compliance. The classic case study of this beneficiary-oriented strategy is, of course, EPA's implementation of the scrubber requirement. When the Carter administration chose to mandate postcombustion cleanup technologies for all coal-fired power plants, regardless of the sulfur content of coal used, it helped create a powerful coalition of beneficiaries—environmentalists and high-sulfur coal producers. This strategy has been widely criticized as inefficient economic and environmental policy (Crandall 1983b), and that coalition has been described as bizarre (Ackerman and Hassler 1981). But, as we noted above, it may be impossible to implement a standard of technical or economic rationality for agency performance when the policy arena is characterized by ambiguous scientific evidence and high economic stakes. Moreover, we should not be surprised when technical controversies generate unexpected coalitions. Examples of this phenomenon abound, as Rosenbaum (1985, 72–73) has observed:

Environmental groups have been adept at forming alliances with almost any interest whose collaboration might be a political advantage. In the environmental division of Politics Makes Strange Bedfellows, for instance, one might ponder the coalition that opposed in 1981 renewed funding of the Clinch River Breeder Reactor project: major environmental organizations such as the Natural Resources Defense Council and the Friends of the Earth were joined by the International Association of Machinists and Aerospace Workers, the Methodist church's General Board of Church and Society, Rural America, the Union of Concerned Scientists, and conservative groups such as the Heritage Foundation and the National Taxpayer's Union. (The latter two organizations, especially, have rarely been accused of vigorous environmental sympathies.)

Acid Rain and Environmental Politics

Acid rain provides another striking example of the salience of politics and the instability of interest-group coalitions, in part because concerns about acid deposition are intimately linked to the air quality debate. In fact, for years acid rain was viewed by all participants as an offshoot of the controversy surrounding the implementation of the Clean Air Act. Thus, the problems and complexities of acid rain were observed as early as the 1972 Stockholm Conference on the Human Environment, but they received very little attention in the United States for several more years. For example, while bureaucratic interest in the topic was growing in a fragmented way, debate over the 1977 Clean Air Act amendments all but ignored the issue. Even when discussed, acid rain still was a subset of broader energy and environmental concerns. Typical of the bureaucratic view of the subject at mid-decade was the following assessment by the Council on Environmental Quality (1977, 198):

A comprehensive national program to assess the acid rain problem needs to be developed. Current SO_2 and NO_2 standards are directed at protecting health in urban or industrial environments. Standards related to acid precipitation are lacking because definitive assessment information does not exist. Much more effort should be devoted to this problem in view of the fact that we are looking to coal as a major energy source. Coal use will mostly increase emissions of sulfur and nitrogen oxides, unless they are controlled. The potential adverse effects and costs to society of acid precipitation deserve prompt attention.

Here, tied to coal combustion's potential threats to public health, was the beginning of a call for greater research and development focusing on acid deposition.

This argument soon began to gain momentum within the executive branch. In 1979, President Carter called acid precipitation "a global environmental problem of the greatest importance" in a message to Congress asking for expanded research and development as well as possible control measures under the CAA. Coupled with the U.S.-Canada memorandum of intent to negotiate an agreement on transboundary air pollution, the Acid Precipitation Act of 1980 represented the first attempt at the federal level to address the issue in a focused manner. This legislation established the Interagency Task Force on Acid Precipitation, charged with "planning and implementing a comprehensive research program to clarify the causes and effects of acid precipitation" (Council on Environmental Quality 1981, 48).

During the Carter administration, acid deposition had been identified as a problem, and research was to be used to facilitate government intervention to resolve it. R&D policy commonly has been used as part of an overall political strategy. And like other types of policy responses, R&D can be used to facilitate relationships with either beneficiaries or cost-bearers. That is, research and development agendas can be defined to further the goals of parties-at-interest by targeting various substantive areas, focusing on particular technological alternatives, or identifying particular hazards or risks. Thus, for the Carter administration, better science was needed, but generating a more complete knowledge base was not to be used to delay policy responses. As Douglas Costle, Carter's EPA administrator, said in 1980, the time had come to "make the transition from research to action" even though it was recognized that knowledge about acid precipitation needed to be augmented (Tobin 1984, 243).

When the Reagan administration came to Washington in January 1981, this orientation changed dramatically. Research uncertainties and gaps were used to justify the decision that any transition to action should be postponed. The new administration raised the banner of "good science" as its rationale for this policy shift. For Ronald Reagan, regulations based on "bad science" had led to "economic tragedies" which his administration was going to correct. As he put it, "We need to review those regulations and bring them in line with all the new scientific knowledge" (Lash et al. 1984, 130–31). In fact, this rationale represented a change in policy from a corporatist to a co-optive bureaucratic strategy. No longer were environmental policies to be constructed to benefit environmentalists and their allies. The new approach would coincide with the interests of industry and other groups opposed to environmental regulation. The strategy is described in detail by Dickson (1984, 262):

Those either opposing new regulations or seeking to have existing ones weakened have done so by challenging their scientific rationality. Demands from industry are made for "proof" and "certainty" of environmental or health damage before regulations are introduced for new technologies. A detailed description of the relationship between "cause" and "effect" is said to be necessary before any action is taken against the former in order to prevent the latter. Corporate lobbyists contend that scientific questions should be answered before any regulatory action is taken; if only partial answers are available, then regulations should be based on scientific judgments. In a growing number of ways, an appeal to "scientific" rather than "political" judgment has become the touchstone separating "good" from "bad" regulation.

Operating from this perspective, the Reagan administration increased the acid rain R&D budget and maintained that more compelling evidence would be needed before any mitigation policies were put into effect.

Over time, however, the accumulation of scientific evidence itself began to undermine this strategy. The Interagency Task Force on Acid Precipitation released a report arguing that manmade sources of pollution probably were the major cause of acid deposition and declining pH in the northeastern United States. Second, a panel established by the White House's Office of Science and Technology Policy (OSTP) to conduct a peer review of the U.S.-Canada MOI Work Group reports concluded that to postpone action while waiting for definitive scientific knowledge might lead to irreversible impacts on the environment. The OSTP panel urged starting with least-cost SO_2 emission reductions (Nierenberg et al. 1984). Third, the NAS found no compelling evidence to support arguments for a nonlinear relationship between SO_4^- deposition and SO_2 emissions for eastern North America as a whole. Taken together, these analyses put tremendous pressure on the Reagan administration to modify its strategy.

As a consequence, since 1983 EPA has found itself in a difficult position. Returning once again to the agency as its administrator, in the wake of Ann Burford's scandal-ridden tenure, William Ruckelshaus faced strong opposition within the Reagan administration to any control actions beyond those already mandated. The Office of Management and Budget, with its director, David Stockman, took the lead in cabinet discussions, and DOE remained adamantly opposed to costly mitigation programs. In addition, the pressures on the administration were reduced to

Acid Rain as a Political Problem 121

some extent because potential cost-bearers remained at least as united as before, while possible beneficiaries of control programs were still somewhat fragmented.

The best example of this was the fate of the Waxman-Sikorski Bill. This piece of legislation (H.R. 3400) appeared to have some prospects of adoption once the reports of the Interagency Task Force, the OSTP panel, and the NAS were released during 1983 and 1984. However, environmentalists and other supporters of a more aggressive acid rain policy failed to mobilize successfully behind the initiative. The situation in early 1984 was described by Rep. Gerry Sikorski (D-Minn.) as follows:

The public pressure was unbearable and was at the point where a decision should have been made on an issue that's been percolating too long. But it's clear that once again the hard-liners have taken over.

I said three months ago we had a 50-50 chance of getting Clean Air Act amendments in the Congress. Now I don't think the likelihood's there, largely because there's been a failure of the constituent groups to coalesce around HR-3400. As of today, not one group has endorsed the bill. What's happened is that the members are willing to vote and resolve this, but the constituent groups are not.

The environmentalists want more tonnage, at any cost. The United Mine Workers have locked themselves in the closet. They're hoping a year will pass and no legislation will be enacted. We clearly have the bill for them, but they won't endorse it. It's the same with the utilities.... We knew the major law firms of these utilities in many cases were recommending they endorse this bill. Then, just as things started to coalesce, the environmental groups, zip, went off in their direction of 12 million tons or bust. Reagan had been under pressure to come up with something because things looked like they could move, and then they saw this thing pull apart and that was it. They saw splintering and said let's just don't move (quoted in Pawlick 1984, 164–65).

Sikorski's comments illustrate the importance of the second major component of bureaucratic politics—interest-group initiatives—since the situation remains largely unchanged.

The framework developed by Chubb (1983) is useful for evaluating this aspect of acid rain politics. Three major patterns of interest-group initiatives emerge. The first is monopsonistic, where a single benefiting group is virtually the sole bureaucratic participant. The second pattern is competitive, where participation is balanced among and between beneficiaries and cost-bearers. The third is monopolistic, where constituency relationships are controlled by cost-bearers.

For environmental politics in general, interest-group initiatives appear to be highly competitive. This is true even though most of the "rules of the game" seem to favor producer interests. Because both the magnitude and distribution of costs are better known than the benefits, successful group representation is made easier for industrial and other private interests. Group size and the size of the stakes in any particular controversy also favor producers; and self-interest itself also may work to the advantage of private groups because "truly disinterested participation to advocate a general public good, while not unheard of, is extremely rare" (Noll and Owen 1983, 41).

The environmental movement in the United States, however, has exhibited substantial strength over a long period of time. Traditional conservation-oriented groups such as the National Wildlife Federation (NWF) and the Sierra Club have been joined by newer, more litigation-oriented advocacy groups such as the Natural Resources Defense Council (NRDC) and the Environmental Defense Fund (EDF). And while interest groups may not be as easily mobilized in anticipation of some vague configuration of benefits as are cost-bearers in reaction to some specific economic burden, once a set of regulations is in place, beneficiaries have a dramatically increased motivation to act. Thus, once the initial legislative and administrative framework for air quality management was established in the early 1970s, powerful interest groups could begin to identify existing benefits and to anticipate their expansion. Since then, the NRDC, EDF, and NWF have been at the forefront in advocating ways to control air pollution.

Moreover, in the environmental protection area, the notion of "benefits" is often broadly conceptualized. Many environmental groups operate as if important social benefits are to be obtained from their efforts to balance the power of corporate and bureaucratic groups (McFarland 1976). A related point is that environmentalists may have redefined the basis for valuing benefits from a social perspective. Environmental protection is a part of what has been labeled "new social regulation," which is apparently largely advocated by a well-educated, relatively affluent professional and managerial class. Advocates may focus on forming more environmentally conscious decision-making frameworks in industry and government or changing income distribution rather than stressing the more traditional economic benefits associated with the "old regulation" of the populist and progressive eras (see Milbrath 1984; Weaver 1978). And while it is extremely difficult to measure social benefits, the environmental movement generally takes as an article of faith that these improvements are worth the costs of compliance (Center for Policy Alternatives 1980). Because there are obvious benefits to society from controlling threats to health and the ecological balance, suggesting that the environmental movement seeks to reorient the economy does not mean that it constitutes a conspiracy. Such advocacy is often found among an educated and affluent class, however, because awareness of and respect for those benefits varies among subgroups of the population, in spite of broad-based support for environmental protection.

Yet, American environmental groups, while increasingly active in the evolving debate over controls, were not very important actors in the initial process of getting acid rain on the international political agenda. Instead, the first major initiatives came from Scandinavia. This has proved to be an important factor in the evolution of interest-group strategies, because the environmental coalition has strong international linkages. In fact, other national governments, and particularly those of Sweden and Canada, have been important allies in articulating the seriousness of the issue. This position has acknowledged the importance of expanding the scientific base for action, but not at the expense of taking action now to reduce emissions. Costs of compliance, according to this argument, should be borne by the polluter. From the perspective of environmentalists, domestic administrative resources for research, monitoring, pollution control and abatement, compliance and enforcement, and assistance to state governments should be increased (Friends of the Earth 1982, 108–11). In addition, they argue that greater efforts toward international cooperation are needed.

124 THE ACID RAIN CONTROVERSY

The Carter administration, except for DOE, was basically supportive of these initiatives. By the end of the Carter years, the initial steps in establishing a framework for coordinating research had been taken, the transboundary aspects of the acid rain problem were the focal point for bilateral discussions with the Canadians, and amending the Clean Air Act to enable EPA to address long-range transport problems was being considered (Council on Environmental Quality 1981).

Under President Reagan, this policy orientation changed. Interest-group initiatives from the industrial community, largely ignored outside of DOE during the Carter years, found a receptive audience in the new administration. Articulated by bodies like the Electric Power Research Institute (EPRI) and the Edison Electric Institute (EEI), this strategy relied upon questioning the scientific underpinnings of government policy. According to EPRI's Ralph Perhac:

I cannot overemphasize the importance of knowing what responsibility the industry has for acid deposition. Unless we know that, we cannot judge the efficacy of any control strategy which might be promulgated and, as scientists, we have an obligation to evaluate the extent to which a control strategy will achieve its goal. For acid precipitation, we cannot do that (Editorial Research Reports, 1982, 69–70).

Utilities and other industrial polluters certainly are not alone in making this case. They have powerful allies in the eastern and midwestern high-sulfur coal industry, the United Mine Workers, and in public officials from eastern and midwestern coal-producing states. The unhappy coincidence that cost-bearers in this case also tend to be located in economically depressed parts of the country gives the argument additional weight. Thus, acid rain contributes not only to transboundary conflicts in the international sense, but also to regional disputes within this country. And the Clean Air Act, with its focus on local air quality, is poorly designed to respond to these conflicts (Russell 1982).

Once the "bad science" argument had been undercut by the reports of the NAS, the OSTP Peer Review Panel, and the Interagency Task Force, those industries that were likely cost-bearers shifted strategies to focus on this macroeconomic problem. And they added the assertion that acid rain controls might damage American industrial competitiveness abroad, especially in areas such as steel production. An explicit case was made by trade associations such as EEI and the National Coal Association that both national and regional economics demanded a cautious approach to the acid precipitation issue.

By 1984, interest-group initiatives were entrenched on both sides of the debate. Environmental groups like NRDC, EDF, and NWF were joined by a range of other beneficiary parties-at-interest. Recreational interests and government officials from affected states, such as New York, Vermont, and Maine, and from the League of Women Voters and other organizations have received support from the Canadian Coalition on Acid Rain and the Canadian government. The most striking examples of the latter were two documentaries produced by the Canadian National Film Board, for distribution in the United States. Rosenbaum (1985, 138) suggests that the political alignments were predictable:

Midwestern industries and utilities, together with their local and congressional political spokesmen, will resist assuming the full burden for pollution abatement, as some legislation already proposes. The coal industry—fearing that regulation may mean a massive cutback in the use of high-sulfur Appalachian coal, if not all coal—will seek to blunt the impact of regulation upon the industry as much as possible. State governments will generally insist upon a national set of standards for any control technologies or ambient air quality goals associated with acid precipitation, to avoid the politically and economically difficult problems they would inherit if too much enforcement discretion remained with them. Most environmental groups, long impatient at the pace of current efforts to control acid rain and snow, are largely committed to a strong national regulatory program. In this they are joined by the Canadian government, which asserts that two-thirds of the acid precipitation falling in its southeastern and midwestern provinces probably originates in the United States.

Compromise in such a situation may be very difficult. It will not be made any easier by the Reagan administration's initial polarization of the entire issue. As Tobin (1984, 247) says, by taking "extreme and politically untenable positions," the administration missed a chance in the early 1980s to revise the Clean Air Act to improve its effectiveness while at the same time reducing its costs.

In this situation, the dominant interest-group pattern has been a form of modified capture of EPA. Interest-group initiatives continue to be presented competitively, with beneficiaries battling cost-bearers for influence in the administrative process, but the cost-bearing end of the constituency continuum appears to have dominated bureaucratic strategy and standard-setting. EPA thus appears to be a classic example of modified capture. As Chubb (1983, 27) has characterized this kind of bureaucratic-clientele relationship, it is "a regulatory agency trying to co-opt its opposition, rather than supporting its supporters, in a competitive constituency."

Implications for Policymaking

With action at EPA stalemated for at least the short term, and with other major bureaucratic actors such as DOI and DOE reflecting producer interests (see Tolchin and Tolchin 1983), the focus of attention has shifted to Congress. However, most of the debate thus far has downplayed how these various proposals might shape interest-group politics in the future. The very different patterns of actor mobilization and interaction that are likely to occur as a result of the allocation of acid rain control benefits and costs are apparently at the core of the debate.

Wilson (1980) provides a useful typology for these anticipated patterns of interaction. He suggests four categories. When both the costs and benefits of a regulatory policy are widely distributed, then majoritarian politics tend to prevail. If both the costs and benefits of an action are narrowly concentrated, the political process is more likely to reflect the dynamics of interest-group politics. Client politics, on the other hand, traditionally has predominated, when benefits are concentrated but costs are distributed widely. In Wilson's terms (1980, 369), client politics are characterized thus:

Some small, easily organized group will benefit and thus has a powerful incentive to organize and lobby; the costs of the benefit are distributed at a low per capita rate over a large number of people, and hence they have little incentive to organize in opposition—if, indeed, they even hear of the policy. . . . Client politics produces regulatory legislation that most nearly approximates the producer-dominated model.

An entrepreneurial style of politics may emerge when benefits are spread broadly but costs are concentrated narrowly. Each of these categories appears to have relevance for characterizing the debate over acid rain controls.

The Limits of Client Politics

Restricting acid rain policy solely to a research and development strategy—the consistent position of the Reagan administration—has become increasingly less defensible as studies of the phenomenon proliferate. As we have seen for some time, environmental groups such as EDF and NRDC, as well as other parties-at-interest such as New England and New York state officials, have been critical of the Reagan administration's policy of requiring more scientific research before taking action, terming it a stalling tactic in the face of significant empirical evidence that a hazard exists (Boyle and Boyle 1983; Louma 1980). As important as the growth of scientific knowledge has been in supporting this criticism (Mosher 1983b), even more significant has been the emerging perception of unacceptable inequities in the current distribution of costs, damages, and benefits. This view has reinforced calls for immediate reductions as well as research.

The well-organized advocates of the status quo are, of course, those firms that have tried to reduce their waste disposal costs by shifting them to the environment and thus to society as a whole through client politics. As noted above, utility systems, primarily located in the Midwest, that generate electricity from high-sulfur fuels are significant actors in this process. Electric utilities are not the only supporters of the status quo, however. High-sulfur coal producers and miners, some railroads, chemical manufacturers, and midwestern public officials are also important (Kahan 1986). Each of these groups, especially high-sulfur coal interests, has a vested interest in opposing acid rain control measures. This is especially true since controls would result in a significant redistribution of costs and benefits.

On the other side, a coalition has formed to press for modifications in the existing situation. The acid rain control lobby is composed of environmentalists, landowners, and small businessmen in areas affected by acid rain (particularly Ontario and New York resort communities), Canadian power interests, and the Canadian government itself (Scott 1986; *Fortune* 1983). In fact, the Canadian parties-at-interest have expanded the definition of client politics to include the actions of the U.S. government. A report of the Canadian House of Commons (1981, 109) declared:

Acid rain is produced because firms, operating in their own best interests, function within an institutional framework which cannot effectively manage the environment. Under some circumstances, the same argument can be made with respect to political jurisdictions. Since acid rain is associated with the long-range transport of air pollutants, emissions operating in one political jurisdiction can be deposited in another jurisdiction. Thus any government which operates in the best interests of its own citizens will tend to do little about controlling emissions which fall in, and cause damage to, other provinces, states, or countries. *Just as acid rain allows firms to impose external costs on third parties, acid rain allows one political jurisdiction to impose such costs on other jurisdictions. In this respect governments behave like private firms.*

To date, the acid rain control lobby has focused on attempts to keep the issue visible and on efforts to influence legislative and executive initiatives. For example, environmental groups have developed their own sources of scientific and technical expertise. But, as is typical of client politics, they have had limited success in organizing broader coalitions among the parties-at-interest who might benefit from a change in the status quo. This is a direct result of the fact that the benefits from changes in the existing system are highly uncertain and widely diffused, while the costs of change are highly concentrated and more certain.

Nonetheless, the anti–acid rain lobby continues to press for additional reductions in SO_2 emissions, and some modifications in the status quo appear quite likely. Attempts to expand the policy framework to include some sort of international cost-sharing appear less promising in the near term. However, there has been some movement even in these areas.

The question becomes when further controls can be mandated, what the timetable for achieving those reductions will be, and how those costs shall be amortized. Two possible scenarios under which an answer may unfold are suggested by the range of financing alternatives available: abandoning client politics in favor of either entrepreneurial or majoritarian political arrangements.

"Polluter-Pays" or Cleanup Subsidies?

Entrepreneurial Politics On the surface, the most obvious way to restructure acid rain policy would seem to involve following the trend of environmental politics in recent years: implementing entrepreneurial arrangements. Wilson (1980, 370) characterizes this option as follows:

A policy may be proposed that will confer general (though perhaps small) benefits at a cost to be borne chiefly by a small segment of society.... Since the incentive to organize is strong for opponents of the policy but weak for the beneficiaries, and since the political system provides many points at which opposition can be registered, it may seem astonishing that regulatory legislation of this sort is ever passed. It is, and with growing frequency in recent years—but it requires the efforts of a skilled entrepreneur who can mobilize latent public sentiment (by revealing a scandal or capitalizing on a crisis), put the opponents of the plan publicly on the defensive (by accusing them of deforming babies or killing motorists), and associate the legislation with widely shared values (clean air, pure water, health and safety).

While environmentalists point to the dying spruce trees on Camel's Hump Mountain in Vermont as an example of ecological crisis, the current scientific uncertainties surrounding the effects of acid rain make it difficult to create widespread awareness of a crisis, as happened with Three Mile Island or Love Canal, although few informed individuals would argue that no problem exists. Similarly, while individual scientists have legitimated acid deposition as a R&D topic, no dominant policy entrepreneur has emerged in the U.S. political context. Without either or both of these factors, the political feasibility of entrepreneurial politics appears to be very low.

This is because abandoning client linkages in favor of entrepreneurial politics involves dramatic changes in how costs and benefits are distributed. Most importantly, it involves the implementation of the economic philosophy that the "polluter pays." As a result, entrepreneurial politics are compatible with using the normal electric utility rate-making process as well as emissions fees or sulfur-content-in-fuel taxes as alternatives to that process. Each would have the effect of imposing the cost of control at the primary source of SO_2 emissions.

While placing the entire burden on the electric utility companies and leaving it to them to allocate the costs among their customers and shareholders may appear attractive for both efficiency and equity reasons, it is also highly controversial. From the point of view of the midwestern states, it is simply too big a financial burden to bear. As a result, interests in those states have pressed for various spread-the-cost schemes.

Nevertheless, some variation of the "polluter-pays" principle has dominated U.S. environmental politics to date. Regional conflict certainly would be exacerbated by any scheme to subsidize cleanup efforts in the Midwest. This is especially true, given the widespread perception in other regions that midwestern interests have not been aggressive enough in reducing their emissions. Edwin Rothchild, of the Citizen/Labor Coalition, is blunt on this matter: "There is no question that these Midwestern utilities have been lax in complying with the Clean Air Act. They have been stalling for years in avoiding the costs of installing scrubbers" (quoted in Mosher 1983a, 999). Moreover, some participants in the debate are concerned that continuing down the technology-forcing path of forced scrubbing represents a bailout of the high-sulfur producers. And, as Ackerman and Hassler (1981) have asserted, there is the possibility that FGD-forcing measures may have the perverse consequence of selectively increasing emissions in the East. Yet, to achieve a reduction of the magnitude proposed by acid rain control advocates of approximately 40 to 50 percent within ten years, it becomes impossible not to rely on a FGD-retrofit strategy.

Certainly, a continuation of the decade-long emphasis on

engineering mechanisms to ensure reliance on the command-and-control approach to implementation of the "polluter-pays" philosophy is likely to lead to entrepreneurial politics linking such seemingly dissimilar interests as the clean air and dirty coal groups (Ackerman and Hassler 1981). This is an important consideration. We know from experience that the command-and-control approach can lead to some very strange, yet very powerful coalitions of energy, environmental, and economic actors. In fact, this highly volatile and unpredictable dimension of entrepreneurial politics is a major reason for the appeal of various spread-the-cost subsidies.

Majoritarian Politics The most widely suggested departure from entrepreneurial politics is to establish some majoritarian political framework. Wilson (1980, 367) describes this orientation: "All or most of society expects to gain; all or most of society expects to pay. Interest groups have little incentive to form around such issues because no small, definable segment of society (an industry, an occupation, a locality) can expect to capture a disproportionate share of the benefits or avoid a disproportionate share of the burdens." This, in a nutshell, is the great appeal of majoritarian policy. Regarding acid rain, such financing options as generation fees, Btu taxes, or direct federal budget outlays promise to spread the costs of mitigation across the broadest possible segment of the society. Each reflects the classic majoritarian strategy for formulating a policy response to a public problem. In an abstract sense, spreading the burden means that the political feasibility of such an option may be greater than more punitive actions based on the "polluter-pays" approach. But majoritarian options are neither as economically efficient nor as equitable as entrepreneurial ones. Thus, while client politics may be fast approaching its limits in the acid deposition case, neither entrepreneurial nor majoritarian politics guarantees consensus.

As an example of the liabilities associated with majoritarian options, consider the following assessment of the use of a generation fee, or "superfund" approach:

An important political and ethical liability of the superfund concept is that it may oblige "victims" to pay the "villains" to stop the offense in question; the antagonist is entitled to economic assistance from the victim. It would amount to a subsidy on the part of ratepayers in relatively nonpolluting states, such as New Hampshire and Vermont (whose electricity rates are already high, in part because state laws and regulations prohibit the use of dirty fossil fuels), to ratepayers whose electricity is generated from the combustion of sulfurous coal (Rhodes 1984, 29).

The Missing Ingredient: Concentrated Costs and Benefits

Like most environmental issues, the acid rain problem does not fit easily into a framework characterized by concentrated costs and benefits. We know substantially less about the benefit side of the acid deposition equation than we do about costs. The value of damages prevented or mitigated is largely speculative. But we can anticipate that benefits will accrue to different parties-at-interest, at different times, and at different magnitudes than is true of costs (Regens and Crocker 1984). And we also know that a strong case can be made that both the U.S. political system and the international system function best when the various participants have an understanding of distributional consequences, that is, how they stand to benefit from or pay for a particular policy. Environmental policymaking is no exception to this rule. Rosenbaum (1977, 21) has made this point a central part of his analysis:

Public policy-makers are very likely to weigh the costs and consequences of environmental policy decisions in group terms. This is not to suggest that various groups are the only components in the public official's subjective map of the political world, but almost all studies of official decision-making emphasize that group viewpoints and the group consequences of policy choices do weigh heavily in official thinking much of the time. It is a matter of fundamental political intelligence for officials to discover, in any case, where important interests stand on policy questions, environmental or otherwise. All this underscores the fact that what groups do in the political arena—how they represent their interests, to which officials they speak—is likely to have substantial consequences in shaping public policy.

This is conventional wisdom. Yet this fact alone may go a long way toward explaining our inability to develop agreement on the elements of, and timetable for, a control program as well as a financing strategy acceptable to the diverse range of participants in the acid rain arena.

Only by more precise specification of winners and losers will it be possible to expand the range of financing options to include more innovative approaches, such as compensation schemes. Transfer payments are certainly not without their problems (see Goldfarb 1980), but some experts believe we might find the real cost of transfers less expensive than current pollution control policies (Russell 1982). Moreover, compensatory programs, at least of an indirect nature, are already at the root of the acid deposition controversy. A superfund of the kind discussed above really represents indirect compensation for polluters, but lacks the economic efficiency and equity advantages of more direct compensatory measures.

Under the guise of regulatory relief, the Reagan administration has attempted to establish environmental policies that favor the cost-bearers of pollution control across the board. In the short run, this approach to regulation has led to relief, albeit often limited, for those sources of pollution and other environmental externalities. But the longer-run consequences of allowing regulatory agencies to establish a symbiotic relationship with, if not to be captured by, cost-bearing interests appear to be heightened conflict and a return to the veto politics of a decade ago.

Perhaps even more discouraging is the lost opportunity for creating new environmental agendas and pollution abatement mechanisms. Acid rain is a problem that will not go away. It creates substantial difficulties in domestic and international relations that need to be addressed. We are not well served by the use of science as a delaying tactic, at best, and as a cloak for ideological positions, at worst. Perhaps most significantly, we are badly served by a rhetoric that acts as if economic calculations of winners and losers are independent of political choices. Reich (1983, 369–70) makes this point:

When political choices have been made, it has been in the guise of economic policy. One example is the debate over the environmental effects of energy development. Decisions to embark on large-scale energy projects have benefitted those who gain access to cheap power while simultaneously imposing losses on citizens who live near nuclear generators or hazardous waste sites or who care deeply about protecting America's beaches from oil spills and its wilderness from strip mining. Yet these political decisions, submerged in the regulatory process, have come increasingly to be based on an abstract tally of costs and benefits—an analytical exercise that cannot engage and thus ignores the distributive issues.

Acid rain control decisions ultimately will be resolved according to precisely this kind of distributive calculus, since they represent explicit political choices. In a democracy, such choices, to be legitimate, should be made within the public rather than the private context.

6
PROSPECTS FOR POLICYMAKING

 The possibility of significant acid rain control awaits the transformation of the political tactics and ideological horizons of opposition forces to environmental degradation beyond environmental doomsaying and interest-group reforms.
—*Ernest Yanarella (1985)*

THE PREVIOUS CHAPTERS have demonstrated graphically that there is a wide spectrum of opinion as to the correct stance to take vis-à-vis acid rain. Developing a consensus on an appropriate policy response is particularly difficult because the acid rain problem has complex environmental, economic, and political facets. Since 1981, each of the opposing sides in the debate over mandating controls has collected a growing stack of scientific studies and economic analyses documenting the point it wants to make. For example, S. William Becker, executive director of the State and Territorial Air Pollution Program Administrators (STAPPA), observed: "We've finally reached a point where science has confirmed what we knew all along—that what goes up comes down, and what comes down is harming many parts of the country. Unfortunately, even the most prestigious [scientific] institutions' reports have a little for everybody. No one scientific report is going to put this debate to an end" (quoted in Stanfield 1986a, 1501). As a result, in the ongoing debate, the scientific information currently available does not lead unequivocally to a conclusion about whether it is appropriate to begin additional control measures now or to await better understanding. This is true of questions about the scope and pace of environmental effects as well as issues surrounding the effectiveness of proposed emissions reduction

135

strategies, both in terms of preventing future damage and reversing damage that already has taken place. Similarly, considerable uncertainty clouds the benefits and costs of any proposed acid rain control policy.

As we have observed, since taking office in January 1981, the Reagan administration has adopted the position that further information is needed before a decision can be made to require additional emissions reductions for acid rain control. Thus, in August 1981, the idea of accelerating the federal government's research program under the aegis of the Interagency Task Force on Acid Precipitation was incorporated into the administration's "principles" for Clean Air Act revision, and funding for acid rain R&D was increased. In his 1984 State of the Union Address, President Reagan emphasized his efforts to double research funding, called for restoring lakes affected by acid deposition, and supported the development of technology such as limestone injection multistage burners to reduce the precursors of acid rain.

The replacement of Ann Burford as EPA administrator with William Ruckelshaus was at least minimal evidence of the administration's willingness to accept acid rain as an environmental problem, even if its magnitude as well as causes remain uncertain. Out of all this came a 1983 EPA proposal, within the context of the administration's cabinet council system, to reduce sulfur dioxide emissions by 3 million tons in a ten-to-twelve-state area, using an emissions tax to finance the reductions. Largely on the basis of objections raised by the Office of Management and Budget and DOE over whether sulfur dioxide reduction programs to control acid rain were justified, President Reagan did not endorse any program for additional reductions beyond those currently required by the CAA.

Opposition to further emissions reductions remains the administration's position in spite of the 1984 recommendation by an advisory panel to the White House's Office of Science and Technology Policy that the least costly sulfur dioxide emissions reductions be undertaken at once. That panel's recommendation was based on a concern that the environment might suffer irreversible damage if no action were taken until scientific knowledge about the acid rain problem was complete (see Nierenberg et al. 1984). Subsequently, while for the first time conceding that acid rain was a serious problem in his March 1986 acceptance of the special envoys' report, President Reagan continued to maintain that further research was necessary before embarking on further controls. Naturally enough, the Reagan administration's position has been endorsed by electric utilities, high-sulfur coal producers, and other industrial groups likely to be the target of new emissions reduction efforts (Magnet 1983).

Because environmental groups, some states, and the Canadian government view the administration's position as a delaying tactic to avoid taking regulatory action, demands for legislation have not waned as the U.S. government's R&D program has expanded. In fact, while the war of the scientists lingers on, it now takes a back seat to the economic and political battles over acid rain. The forces that gradually have been moving the issue to the forefront, as well as those resisting its emergence from the policy backwaters, approach the fray from across the spectrum.

Action in Congress

Legislation proposing substantial reductions in sulfur dioxide emissions has been introduced in each session of Congress since 1981. Basically, each proposal answers the question— Does present knowledge, given its gaps, provide sufficient information to design a control program?—in the affirmative. Senator George Mitchell (D-Maine), a leading advocate of acid rain controls, has framed the issue along the following lines:

I emphasize the high risk of inaction. I recognize the fact that some questions about acid precipitation remain to be answered. But I also believe that evidence to support a meaningful sulfur reduction is stronger and more complete than in most legislative debates. (U.S. Senate, Committee on Environment and Public Works 1982, 124)

On the other hand, opponents of mandating the emissions reductions approach, not surprisingly, arrive at a different conclusion. For example, Senator Richard Lugar (R-Ind.) has argued that "there is an acid rain problem but no acid rain crisis" (U.S. Senate, Committee on Energy and Natural Resources 1982, 158).

Proponents of controls disagree, however, over how to allocate the costs of any emissions reduction strategy. As Senator John Glenn (D-Ohio) has noted, "The crux of the acid rain cleanup problem has always been the cost of cleanup and who should bear it" (Mosher 1983a, 998). Thus, the issue of financing, including questions of equity, increasingly has become the focus of both the public policy debate and the symbolic politics surrounding allocating the costs of an acid rain control program. The two approaches for addressing the question of acid rain controls illustrate this point. The first essentially embodies the "polluter-pays" principle, while the second, in part as an acknowledgment of the large-scale costs, includes a subsidy approach. Key bills that were first introduced in 1984 reflect these approaches. Those bills continue to frame the ongoing terms of legislative debate.

On March 7, 1984, the Senate Environment and Public Works Committee voted 14–2 in favor of an acid rain bill (S. 768), as part of legislation containing comprehensive CAA amendments, requiring a 10 million ton reduction in sulfur dioxide emissions by 1994 from a thirty-one-state area adjacent to and east of the Mississippi River. In 1980, annual sulfur dioxide emissions from this area totaled about 17 million tons (U.S.-Canada Work Group 3B 1982). Action on S.768, however, was not pursued by the full Senate before the Ninety-eighth Congress ended. The Senate bill, had it been enacted, would have relied on the utility rate-making process to finance acid deposition controls. This approach has continued to be the dominant model for proposals introduced in the Senate in subsequent sessions of Congress. For example, during the Ninety-ninth Congress, Senator Robert Stafford (R-Vt.) introduced S. 2203, the New Clean Air Act, to require substantial reductions as part of an overall revision of emission controls for stationary and mobile sources of air pollution.

The alternative model involving a subsidy for cost-sharing to reduce the economic impact on affected industries, especially those in the Midwest, has had a more receptive audience in the House. While a stalemate has existed since 1981 in the House Committee on Energy and Commerce over proposed acid rain controls, the chairman of its Subcommittee on Health and the Environment, Rep. Henry Waxman (D-Calif.), has consistently introduced bills reflecting this approach to control acid rain. For example, his bill, H.R. 3400, cosponsored by Rep. Gerry Sikorski (D-Minn.), sought to achieve a 10 million ton reduction of sulfur dioxide emissions in the forty-eight coterminous states by January 1, 1993.

The Waxman-Sikorski Bill would have concentrated the bulk of the reduction in SO_2 emissions on fifty electric utility power plants. Power plants having the highest sulfur dioxide emissions in 1980 were singled out for controls. The bill would have required those plants to install FGD systems on a retrofit basis by 1990 in order to reduce substantially their emissions. The top fifty SO_2 emitting electric power plants are located primarily in Illinois (four), Indiana (seven), Kentucky (four), Missouri (five), Ohio (eight), Pennsylvania (four), Tennessee (five), and West Virginia (four). Those plants had the greatest tonnage of sulfur dioxide emissions from among the power plants emitting sulfur dioxide at a rate of three pounds per million Btu's (see U.S. Office of Technology Assessment 1983). Moreover, if one includes an examination of the top one hundred SO_2 emitting power plants, it becomes even more apparent that the largest emitters are concentrated in the Midwest and consist primarily of units that first began generating electricity in the 1950s or 1960s (see Appendix A).

Instead of opting for a "polluter pays" financing approach, with the market essentially reflecting control costs through increases in the price for polluters' products, the House bill would have established a one mill per kilowatt hour fee on all nonnuclear electricity generation in the lower forty-eight states starting in 1985 to fund 90 percent of the capital costs of installing scrubbers on the fifty specified plants. By a 10–9 vote in mid-1984, the subcommittee failed to approve H.R. 3400, which effectively ended any possibility of further control measures being adopted during the Ninety-eighth Congress. Nonetheless, this approach involving cost-sharing was reintroduced with minor modifications as H.R. 4567 in the Ninety-ninth Congress. That bill, however, did not go beyond the hearing stage but reemerged essentially unchanged as H.R. 2666 in the Hundredth Congress.[1]

Because support for the various proposals to control acid rain is mixed in the full Senate and House (see Freeman 1983;

Maraniss 1984), the ultimate form such legislation takes depends on resolving regional disagreement not only about the nature of the problem and the timetable for implementing any program, but also about how to allocate the costs of control. This later point is especially important for several reasons, since variations of both approaches have been reintroduced in Congress. First, both the "polluter-pays" and the cost-sharing approaches to acid rain control concentrate the proposed sulfur dioxide emissions reductions in eight states, located primarily in the Ohio River Valley. Indiana, Illinois, Kentucky, Ohio, Pennsylvania, Tennessee, Virginia, and West Virginia would bear the brunt of the reductions. Those eight midwestern and Appalachian states also happen to be the major producers of high-sulfur coals and account for a majority of U.S. coal production. As a result, key actors in creating the clean coal/dirty air coalition, which promoted the 1977 CAA amendments, would be the major target for sulfur dioxide emissions reductions under either acid rain control strategy.

While the cost sharing approach attempts to reduce the economic impact on those states by forcing the use of scrubbers to protect local coal use and by establishing a national fee to offset some of the control costs, neither approach offers enough to different regional interests. The Midwest would bear the brunt of reductions under either approach because of the high level of emissions concentrated in those states. The version of cost-sharing in various legislative proposals merely reduces the size of rate increases but does not avoid them for midwestern states. New England avoids substantial emissions reductions under either approach. The northeastern states naturally are reluctant to pay to reduce air pollution from midwestern sources, especially since those northeastern states perceive themselves as being victims of that pollution. The "official" southern point of view on national acid rain policy is basically a "research only" position. This view was expressed at the 1983 meeting of the Southern Governors' Association. It stems from several factors. First, emissions from the South do not appear to be as significant contributors to northeastern and Canadian damage as are emissions from midwestern sources. Second, much of the nation's high-sulfur coal is mined in the South, and bills like S. 768 permitting switching to lower-sulfur coals could reduce growth in the southern share of the coal market. Third, average emission rates are lower in the South and West than they are in older industrial regions. But a generation fee like that proposed by H.R. 3400 or H.R. 4567 considers only electricity, not pollution levels. The West shares some of the concerns of the South, especially about any cost-sharing proposal that would continue to restrict the use of western coal while requiring the West to help finance a reduction in sulfur dioxide emissions in the East.

As a consequence, it becomes easier to understand why, in a February 1984 position statement, the National Governors' Association (NGA) could agree to the goal of a major reduction in sulfur dioxide emissions below 1980 levels within thirteen years after the enactment of legislation but could not agree on the specifics of who should pay the cost of control (Sununu 1985). Similar controversy was sparked at its February 1985 meeting. At that time the NGA endorsed legislation introduced by Senator William Proxmire (D-Wis.) and Senator Gordon Humphrey (R-N.H.) that would have required a 10 million ton reduction in sulfur dioxide emissions from the thirty-one-state region east of the Mississippi River over a two-phased, thirteen-year period, financed on a "polluter-pays" basis (S. 503). Governor Richard Celeste (R-Ohio) and other midwestern governors objected to the omission of cost-sharing provisions. While other bills such as the proposal by Rep. Silvio Conte (R-Mass.) also included a phased reduction, with the second phase contingent on findings from the research program, they typically have opted for cost sharing (H.R. 1030). However, without agreement on an environmentally effective and/or equitably framed program, strategies designed to protect high-sulfur coal interests at the expense of economic efficiency or to shift control costs significantly reduce the likelihood of a congressional coalition capable of enacting legislation.

Initiatives at the State Level

But does new congressional legislation offer the only option for addressing the acid deposition problem? The states themselves can take unilateral, albeit limited, steps to reduce emissions from sources within their boundaries. Some states appear to have moved in that direction since 1984. For example, New York's Acid Deposition Control Act was the first state-level legislation specifically intended to curb acid rain precursors. Although the act does not specify how New York industries are to achieve reductions, it mandates the state's Department of Environmental Conservation to develop a plan to achieve a 12 percent reduction from 1980 levels by 1988 and a 30 percent reduction by 1991. In 1985, New Hampshire became the second state to enact an acid rain control law. Its legislation requires a 50 percent reduction in SO_2 emissions to be achieved in two phases, the first involving a 25 percent reduction in emissions by 1990. Wisconsin also has emerged as a leader in state-mandated control programs. The Wisconsin law adopted in 1986 involves a single-phase reduction. The state's five major utilities, which produce over 70 percent of Wisconsin's total SO_2 emissions, are required to reduce their emissions by the mid-1990s to 1.2 pounds of SO_2 per million Btu's of heat input. Sulfur dioxide targets also have been set for all large point sources emitting more than 1,000 tons of SO_2 per year, and NO_x emissions would be capped.

In addition to proposals to unilaterally cap or roll back emissions, states have considered other options. These measures have included highly symbolic steps such as resolutions calling for various responses by the federal government as well as initiating state R&D programs. In its 1984 session, for example, the Massachusetts legislature considered five bills relating to acid precipitation. Proposals ranged from withholding the federal income tax on state employee salaries from the U.S. government until national controls were passed, to placing a ceiling or cap on current emissions or requiring emissions reductions commonly called a rollback. The Massachusetts Department of Environmental Quality Engineering favored a cap because its studies maintain that the state's industries are responsible for only 10 to 30 percent of the acid rain in Massachusetts. Ultimately, the Massachusetts legislature failed to adopt either a cap or a rollback, and instead included $500,000 in its 1984–1985 budget for research on water supply issues, materials damage, terrestrial effects, and monitoring and modeling.

Other states have followed this pattern of emphasizing R&D rather than either caps or rollbacks. For example, by a

Prospects for Policymaking   143

voice vote on February 6, 1984, the Indiana House of Representatives approved a resolution calling on Congress to provide additional research funds to study acid rain. The Indiana legislators also endorsed targeting local sources in the Northeast for emissions reductions and providing national cost-sharing to protect Indiana coal miners. Similarly, Kentucky embarked on a five-year study of acid precipitation to be conducted by the Kentucky Energy Cabinet, the Kentucky Natural Resources and Environmental Protection Cabinet, and the Kentucky Department of Parks. Maine also passed legislation in 1985 initiating a two-year research program. The Maine effort focuses on preparing an NO_x emissions inventory for the state, evaluating the contributions of nitrogen oxides to acid deposition, identifying receptor areas sensitive to acid rain impacts, and estimating the contribution of in-state and out-of-state sources to acid deposition in Maine. However, proposals that would have capped emissions or reduced the amount of sulfur in fuel did not pass the Maine legislature.

Still other states have reached bilateral agreements with Canadian provinces to deal with some of the issues surrounding transboundary air pollution. For example, Michigan and Ontario have agreed to share air quality and acid deposition data, exchange information on air quality standards and trends, and provide each other with annual inventories of emissions and control requirements. In addition, they have agreed to cooperate in joint studies of applying models to estimate dose-response relationships for sensitive ecosystems and to evaluate regulatory strategies.

Taken together, these initiatives, as well as responses by other states, can make little more than a marginal impact on the acid rain problem. For example, responding to constituent pressures to hold down electricity rate increases, the New York State Senate passed a bill in early 1987 to exempt power plants in Nassau and Suffolk counties on Long Island from the state's Acid Deposition Control Act and allow those facilities to use fossil fuel with up to 2.8 percent sulfur content. And even if some states choose individually to cap or roll back their emissions from local sources within their boundaries, such actions promise limited results because acid rain's precursors come from distant as well as local sources. This limitation is implicitly recognized in EPA's State Acid Rain Program (STAR), created in 1985 to work with states on technical and administrative problems that might hinder implementation if a national acid rain control program were adopted. Congress provided $3 million in state and local grants for the STAR program under section 105 of the CAA. As table 17 reveals, most of the states or groups of state and local governments such as STAPPA receiving funding under the STAR program have focused on developing emissions inventories or evaluating the economic implications of control strategies.

Reliance on the Clean Air Act

What then about reliance on the section 115 and section 126 provisions of the Clean Air Act? On December 23, 1983, Senator George Mitchell (D-Maine) sent letters to EPA administrator Douglas Costle and Secretary of State Edmund Muskie calling their attention to amendments of the Canadian Clean Air Act requiring Canada to control emissions of pollutants affecting another country. He also reviewed reports by the International Joint Commission and U.S.-Canada Research Consultation Group on Long-Range Transboundary Air Pollution citing acid rain as a transboundary problem, and requested taking action under section 115 to reduce U.S. emissions. On January 15, 1981, just before Reagan's inauguration, EPA administrator Costle responded to Senator Mitchell's request. Costle found that section 115 of the CAA could be activated and instructed EPA staff to begin work to identify which states should receive formal notice to revise their SIPs in order to control acid deposition (Costle 1981a; 1981b). After the Carter administration left office, however, EPA suspended the activation of section 115.

Events since 1984 clarifying the limitations of those provisions suggest that existing statutory authority does not compel action by the federal government. In 1980 and 1981, New York, Maine, and Pennsylvania filed petitions with EPA that claimed violations of NAAQS within their boundaries due to midwestern emissions. On January 13, 1984, the attorneys general of New York and Maine notified EPA that New York, Maine, Rhode Island, and Vermont intended to file suit in federal court to force the agency to take action on the petitions unless they

TABLE 17
Grants Under the EPA State Acid Rain Program

	Project	Amount Awarded
Alabama	Economic feasibility of emission control techniques	$60,000
Arkansas	Estimating emissions; evaluating pollution control equipment	80,000
California	Acid rain controls and air quality programs	100,000
Connecticut	Statewide control strategy development	50,000
Florida	Economic simulation model of emission control	125,000
Illinois	Emissions inventory; emissions reduction strategy development	140,000
Kentucky	Evaluating SO_2 control techniques	100,000
Maryland	Evaluating emissions caps	74,435
Massachusetts	Statewide emissions control planning; examining interstate emissions trading	227,400
Metropolitan Washington, D.C., Council of Governments	Estimating mobile and area source emissions	50,000
Missouri	SO_2 control for large sources	40,000
Northeast States for Coordinating Air Use Management	Evaluating control technology; examining emissions offsets and caps	209,600
New Hampshire	Examining strategies to achieve emission reduction targets	40,000
New York	Developing cost model for SO_2 reductions; emissions inventory management system	150,000
North Carolina	Evaluating alternative techniques for maintaining an emissions ceiling; revision of PSD program; new source review	100,000
Ohio	Examining techniques to estimate SO_2 control costs for individual power plants	50,000
Pennsylvania	Completion of emissions inventories	120,000
Tennessee, Alabama, Kentucky, and TVA	Examining methods of controlling emissions from large sources; evaluating emissions caps	150,000
Vermont	Maintaining a fixed emissions cap	25,000
State and Territorial Air Pollution Program Administrators	Defining acid rain problem in the West; evaluating impact of national control program on the western states	100,000
Wisconsin, Michigan, and Minnesota	Modeling acid rain control program, evaluating interstate emission controls	230,000

Source: Data provided by EPA in personal communication, R. Brenner, February 1986.

were granted within sixty days. Three national environmental groups—the National Wildlife Federation, Natural Resources Defense Council, and Sierra Club—as well as several individuals, joined in the action to force eight midwestern states to reduce their sulfur dioxide emissions. Litigation was initiated in March 1984.

At an August 1984 hearing before the U.S. District Court of the District of Columbia, EPA stated that while the agency was ready to issue a proposed ruling on the petitions, it was under no obligation to be placed on a timetable to make a final decision. The EPA position was rejected on October 5, 1984, and the agency was ordered to issue a final ruling within sixty days. On December 5, EPA formally denied the petition. The agency asserted that it could provide relief under section 126 of the CAA only under certain conditions and that the act, as written, simply does not address acid rain or long-range transport effects on visibility. The next day, seven northeastern states—Connecticut, Maine, Massachusetts, New Hampshire, New Jersey, New York, and Vermont—appealed the decision to the U.S. Court of Appeals for the District of Columbia Circuit. On July 26, 1985, the court ruled that EPA had to begin the process of controlling emissions thought to cause acid rain in Canada. It gave the agency nine months to impose emissions reductions through the SIP process.

EPA successfully appealed this decision on September 24, 1986, arguing that there was insufficient time for such a plan and that judicial directives should not determine foreign policy (New York v. Thomas, 22 ERC 2241). A three-judge panel of the appeals court unanimously overturned the district court's ruling that the Costle letters did invoke section 115 of the CAA. Writing for the court, Judge Antonin Scalia pointed out that no advance notice of Costle's actions was given, no comments were solicited, and the *Federal Register* failed to publish either the letters or the findings upon which Costle concluded that "acid deposition is endangering public welfare in the U.S. and Canada and [that] U.S. and Canadian sources contribute to the problem not only in the country where they are located but also in the neighboring country" (Costle 1981a; 1981b). As a result, the court of appeals held that the letters did not constitute a rulemaking that could serve as a basis for judicial relief. Because the Supreme Court affirmed the appeals court ruling in mid-1987, the plaintiffs are likely to petition EPA to prepare a formal finding about whether acid rain originating in the United States or Canada is adversely affecting the other country.

Many key participants anticipated that the EPA position would be ultimately sustained. For example, when Senator Robert Stafford (R-Vt.) reintroduced S. 768 on January 3, 1985, as the Acid Rain Control Act (S. 52), he included a provision amending the CAA to make section 126 explicitly address acid rain. Thus, given the seeming difficulty of invoking section 115 or section 126 under current interpretations of the CAA, it appears that new legislative authority is necessary before substantial reductions could be required.

In the interim, EPA's stack-height policy under section 123, now being developed in response to a court order, may achieve some sulfur dioxide reductions on a facility-by-facility basis (Sierra Club v. Ruckelshaus, 19 ERC 1987). If the policy is implemented, EPA would permit compliance with stack-height regulations by emission balancing so that sources required to reduce emissions could contract temporarily or permanently with other sources to make the reductions (50 *Federal Register* 52418 December 23, 1985).

The International Dimension

Finally, a brief review of developments in the international arena reveals a growing consensus that acid deposition is an environmental problem. Acid rain's salience as a transboundary issue with substantial international political and economic implications is underscored by the observation that "acid rain is recognized as one of the two most serious global environmental problems associated with fossil fuel combustion, the other being the accumulation of carbon dioxide in the atmosphere" (Council on Environmental Quality 1979, 70). But, as with responses to the carbon dioxide buildup problem, agreement on effective and equitable regulatory strategies remains elusive.

In the years since the 1972 Stockholm Conference, acid rain's emergence and growing prominence as an issue has produced attempts to address it in a variety of forums. OECD has adopted recommendations for national SO_2 emissions control programs and has endorsed attempts to reduce transboundary air pollution (MacNeill 1983). Parallel efforts were initiated during the mid-1970s in the United Nations Economic Commission for Europe (ECE). The ECE discussions culminated in 1979 with the promulgation of the Convention on Long-Range Transboundary Air Pollution (see Appendix B). The convention provides a basis for voluntary multilateral cooperation to deal with acid rain and other transboundary air pollution problems. The convention offers a potentially useful framework for addressing the acid rain problem, especially air pollution originating in Eastern Europe, because the ECE encompasses the Soviet Union and other Eastern European countries as well as Western Europe, the United States, and Canada. It formally entered into force in 1983 after receiving ratification by twenty-four of the thirty-four ECE member countries.

Both the OECD and ECE endeavors reflect acceptance of the "polluter-pays" principle as a basis for policy action. That concept of international law emerged from the famous Trail Smelter Case which involved air pollution from Canada crossing the border into the United States (United States v. Canada, 3 R. Int'l Arb. Awards 1985 [1949]). It was reaffirmed at the 1972 Stockholm Conference. At that meeting, the Declaration on the Human Environment incorporated the "polluter-pays" concept into principle 21, which states that nations bear a responsibility to assure that their actions do not damage foreign environments:

States have, in accordance with the Charter of the United Nations and the principles of international law, the sovereign right to exploit their own resources pursuant to their own environmental policies, and the responsibility to ensure that activities within their jurisdiction or control do not cause damage to the environment of other States or of areas beyond the limits of national jurisdiction.

However, OECD and ECE rely on voluntary international cooperation to achieve compliance with any emissions targets. Presumably, such reductions would be attained through the use of the best available technology that is economically feasible (see Wetstone and Rosencranz 1983; Swedish Ministry of Agriculture 1982).

In North America, the United States and Canada have made limited progress in developing a bilateral agreement on transboundary air pollution. In 1978, the two governments established a Bilateral Research Consultation Group on the Long-Range Transport of Air Pollutants to coordinate the exchange of scientific information on acid deposition (Altshuller and McBean 1979). During the fall of 1978, Congress passed a resolution calling for bilateral discussions to preserve and protect the two nation's mutual air resources. On August 5, 1980, the two governments signed a Memorandum of Intent (MOI) concerning Transboundary Air Pollution as a framework for bilateral negotiations (see Appendix C). Formal negotiations started in fall 1981. Discussions have continued since then, and acid rain has been included as a topic for conversation at the bilateral meetings between President Reagan and Canadian prime ministers Trudeau and Mulroney. While both governments have acknowledged that acid rain is a serious problem, especially as it affects diplomatic relations, those negotiations continue to be marked by disagreement over a 1982 Canadian proposal for a joint 50 percent reduction in SO_2 emissions. U.S. officials characterized the idea as premature and instead urged continued cooperation under the auspices of the MOI to enhance scientific understanding of the phenomenon (see Marshall 1983; Editorial Research Reports 1982, 63–80).

As a consequence, countries such as Canada, the Federal Republic of Germany, and Sweden, which advocate emissions reductions beyond those currently mandated to meet air quality management requirements, have adopted a dual strategy. First, their approach relies on domestic controls of varying stringency. For example, the planned national reduction goal for the Federal Republic of Germany involves a decline in annual sulfur dioxide emissions from a 1980 level of approximately 3.5 million metric tons to 2.3 million metric tons within the next ten years. The specific regulations to accomplish such a reduction focus on controls of major sources, primarily power plants. Canada, too, has moved recently to reduce emissions. While the United States and Canada have not concluded a bilateral agreement to control transboundary air pollution, especially acid deposition, the Canadian government initially declared that it would seek to reduce emissions unilaterally by 25 percent. On March 6, 1984, Canada set a new reduction goal of 50 percent below 1980 levels by 1994 (Marshall 1984). A year later the Canadian federal government announced additional measures designed to toughen automobile emission standards, develop new control technologies, expand R&D, and create an Acid Rain Office (Environment Canada 1985). As noted above, the federal government and the governments of seven eastern provinces and affected industries have agreed to share costs. Although agreement was reached to share the expense of accomplishing these cutbacks, the Canadian federal government and the provincial governments of Manitoba, New Brunswick, Newfoundland, Nova Scotia, Prince Edward Island, Ontario, and Québec did not establish a specific mechanism to distribute these costs among themselves and affected industries.

The second element of this strategy involves efforts to encourage multilateral agreement on specific reductions of sulfur dioxide and nitrogen oxide emissions under the aegis of the ECE Convention on Long-Range Transboundary Air Pollution. The 1982 Stockholm Conference served as an initial forum for such an attempt. Although the Nordic countries and Canada were unsuccessful in getting agreement on specific target loadings of acidic inputs to protect aquatic ecosystems or level of emissions reduction, the Stockholm Conference marked the first major international meeting at which the Federal Republic of Germany joined with other Western European countries and Canada and formally endorsed instituting additional acid rain control measures. By the June 1983 meeting of the convention's Executive Body,[2] the idea of a 30 percent reduction in sulfur dioxide emissions over a ten-year period had been endorsed as a draft decision, with the United States and the United Kingdom the only major Western countries abstaining from the emerging consensus.

Subsequently, the environment ministers of Canada, West Germany, and eight other Western European countries signed a declaration in Ottawa on March 21, 1984, to reduce sulfur dioxide emissions at least 30 percent from their respective 1980 levels in the coming decade (see Appendix D). Austria, Denmark, Finland, France, the Netherlands, Norway, Sweden, and

Switzerland joined Canada and West Germany as the original members of what quickly came to be labeled the "30 Percent Club." This commitment was a highly symbolic means with which to pressure neighboring countries—especially the United States, the United Kingdom, and Belgium—to make similar pledges.

In June 1984, West Germany, in cooperation with the executive secretary of ECE, sponsored a multilateral conference focusing on aquatic and forest damages from air pollution, especially acid deposition. The underlying objective of the conference, held in Munich, was to set the agenda for the September 1984 meeting in Geneva of the ECE's Executive Body. Just as the 1982 Stockholm meeting helped to build support for obtaining the necessary number of countries to ratify the convention and to bring it into force by 1983, the Ottawa and Munich meetings two years later established a time frame—by 1993—and a target—at least 30 percent—for a specific agreement on reducing annual national sulfur dioxide emissions or of transboundary fluxes. The declaration also commits those countries to efforts to reduce associated co-pollutants, particularly nitrogen oxides.

Although the United States and the United Kingdom declined to endorse such a target, the September 1984 meeting did, in fact, produce an agreement that follow-up discussions would be held on incorporating the 30 percent cut into the legal framework of the convention. Moreover, at that meeting, an additional eight countries agreed to reduce their SO_2 emissions or transboundary fluxes by 30 percent by 1993. Belgium, Luxemborg, Lichtenstein, Italy, the Soviet Union, East Germany, Bulgaria, and Czechoslovakia joined the original members of the 30 Percent Club in endorsing the reduction. And a working group was established at the 1984 Executive Body meeting in Geneva to attempt to draft a protocol for SO_2 control within the context of the 30 percent reduction concept. ECE also has undertaken parallel work to draft a protocol for the reduction of nitrogen oxides. Because ECE relies on voluntary international cooperation to achieve any emissions targets, any willingness actually to reduce or cap emissions will reflect a calculus of the costs and benefits of controlling SO_2 and/or NO_x emissions at the national level for individual countries. The growing support on a multilateral basis for emissions controls, however, has become an increasingly important element in the evolution of the acid rain controversy.

Symbolically, at least, this series of international developments may reinforce the arguments for taking further control actions now. This is reflected in the March 1986 acceptance by President Reagan of the U.S.-Canada special envoys' recommendation for increased emphasis on the development of clean coal technologies to abate acid rain (Stanfield 1986). The United Kingdom took an even more dramatic step at the meeting of European Economic Community (EEC) environment ministers by endorsing the 30 percent reduction concept. Britain's proposal called for a 30 percent reduction in SO_2 output by 1993, leading to 45 percent less SO_2 emissions in the year 2005. Although it was rejected at the EEC meeting after Spain and Ireland refused to accept the new limit, the United Kingdom's proposal represented a dramatic move toward symbolic consensus on the question of acid rain (Dickson 1986). As a result, the policy stance of the Reagan administration has become more isolated internationally.

Where Do We Go from Here?

More than six years have passed since Kathleen M. Bennett, then the Reagan administration's EPA assistant administrator for air, noise and radiation, asserted in a February 1982 hearing of the Senate Environment and Public Works Committee that the scientific knowledge base for policymaking would be much stronger in three to five years: "Within two to three years, we expect to have some very important findings that will help guide us. It is possible that those will suggest further activities that are necessary, and it is possible they won't" (quoted in Mosher 1982, 457). Yet the controversy over acid rain continues to be a major irritant in U.S.-Canadian relations as well as a source of acrimonious intersectoral conflict in the United States.

Bowing in small measures to such concern, especially from the current conservative government of Canada, President Reagan endorsed the report by Drew Lewis, his former transportation secretary, and William Davis, the Canadian special envoy on acid rain, which called for a five-year, $5 billion commit-

Prospects for Policymaking 153

ment by the United States to test new ways of reducing industrial SO_2 emissions, mostly from coal-fired power plants in the Midwest. Half the money would come from the federal government and half from private industry. In fact, although it was authorized by Congress before the report was delivered to President Reagan and Canadian Prime Minister Brian Mulroney on January 8, 1986, there was already $400 million earmarked specifically for clean coal technology. Actually, much of the funding was approved by Congress over initial Reagan administration objections, since the administration had requested only $85 million for FY 1986. Moreover, the United States' reluctance to commit itself to an aggressive program of emissions reductions is underscored by the president's resistance to blaming industry for the problem and authorizing large-scale federal funding. Moreover, he continues to believe that "serious scientific and economic problems remain to be solved" before the acid rain problem can be removed from the bilateral agenda (Office of the Press Secretary, The White House 1986, 1). As a result, any initiative on acid rain originating in the executive branch is much more likely to emphasize symbolic rather than substantive actions, especially for the near term (see Stanfield 1986b).

What, then, of action by Congress? A bipartisan bill, H.R. 4567, that would have limited significantly SO_2 and NO_x emissions, was introduced in the House by Rep. Sikorski (D-Minn.) on April 10, 1986. The House Subcommittee on Health and Environment approved the measure in May, but the Ninety-ninth Congress ended without the bill being considered by the full Committee on Energy and Commerce. The Acid Deposition Control Act of 1986 was cosponsored by 151 House members representing virtually every region. It incorporated several compromises in order to achieve a 10 million ton reduction in SO_2 emissions and 4 million ton reduction in NO_x emissions. Each state would be required to meet only an average, statewide standard for SO_2 and NO_x emissions. The bill would set the emissions standard at no more than two pounds per million Btu's of heat produced per month by all coal-burning electric power plants in each state. That standard, however, would not have to be met until January 1993.

Unlike Waxman's and Sikorski's previous bill, H.R. 4567

did not target the fifty dirtiest electrical utility power plants in the Midwest by requiring the installation of scrubbers. The bill, while forcing utility companies to pay for controlling their emissions, would have allowed the power companies to pass through to consumers a portion of the costs of compliance with the proposed federal standard. Such reliance on the market to embody the "polluter-pays" principle, because it is redistributive in nature, might generate opposition due to the potential increase in the cost of the polluter's products. Any increase in electricity rates, however, would have been limited to not more than 10 percent. Moreover, H.R. 4567 placed the burden on the states to develop strategies for meeting their targets. Such an approach offers the potential for breaking the current impasse and is similar in concept to the existing SIP system.

But merely introducing legislation, even with sponsorship by approximately one-third of the House, is only the first step in what is likely to be a lengthy process prior to actual adoption. There are clear divisions in support for such a regulatory approach to the acid rain issue along both regional and partisan lines. Although 34.7 percent of the total membership of the House of Representatives co-sponsored H.R. 4567, table 18 reveals that such support was concentrated primarily in delegations from eastern states; 53 percent of the sponsors were from the East, 27.8 percent from the West, 15.2 percent from the Midwest, and only 4 percent from the South.[3] In fact, while 71.5 percent of all sponsors came from the states adjacent to or east of the Mississippi River, which would shoulder the brunt of emissions reductions, only 35.8 percent of all members from that thirty-one state region endorsed H.R. 4567. A similar pattern emerges when one examines support based on party identification. Co-sponsors are overwhelmingly Democrats (71.5 percent) as opposed to Republicans. This split partially reflects reluctance by many Republican members to embrace acid rain control, given the Reagan administration's opposition to such measures. The more dramatic cleavage, however, occurs along regional economic lines.

The outlines of two coalitions emerge, with westerners who might benefit potentially from coal market shifts or who lack strong constituent opposition to acid rain controls joining representatives from the East to support H.R. 4567. Opponents, on

H.R. 4567 Sponsorship by State
(in percent)

	In State Delegation	Democrats in Delegation[a]	Republicans in Delegation[b]		In State Delegation	Democrats in Delegation[a]	Republicans in Delegation[b]
Alabama	0.0	0.0	0.0	Montana	0.0	0.0	0.0
Alaska	0.0	—	0.0	Nebraska	0.0	—	0.0
Arizona	20.0	100.0	0.0	Nevada	33.3	100.0	0.0
Arkansas	0.0	0.0	0.0	New Hampshire	100.0	—	100.0
California	57.8	92.6	5.6	New Jersey	100.0	100.0	100.0
Colorado	33.3	100.0	0.0	New Mexico	33.3	100.0	0.0
Connecticut	100.0	100.0	100.0	New York	88.2	89.5	86.7
Delaware	0.0	0.0	—	North Carolina	0.0	0.0	0.0
Florida	21.1	25.0	14.3	North Dakota	0.0	0.0	—
Georgia	10.0	0.0	50.0	Ohio	4.8	9.1	0.0
Hawaii	100.0	100.0	—	Oklahoma	16.7	20.0	0.0
Idaho	0.0	0.0	0.0	Oregon	20.0	33.3	0.0
Illinois	13.6	23.1	0.0	Pennsylvania	39.1	30.8	50.0
Indiana	0.0	0.0	0.0	Rhode Island	100.0	100.0	100.0
Iowa	50.0	50.0	50.0	South Carolina	0.0	0.0	0.0
Kansas	20.0	50.0	0.0	South Dakota	100.0	100.0	—
Kentucky	0.0	0.0	0.0	Tennessee	0.0	0.0	0.0
Louisiana	12.5	16.7	0.0	Texas[c]	14.8	25.0	0.0
Maine	50.0	—	50.0	Utah	0.0	—	0.0
Maryland	50.0	66.7	0.0	Vermont	100.0	—	100.0
Massachusetts	90.9	90.0	100.0	Virginia	0.0	0.0	0.0
Michigan	11.1	18.2	0.0	Washington	25.0	20.0	33.3
Minnesota	87.5	100.0	66.7	West Virginia	0.0	0.0	—
Mississippi	0.0	0.0	0.0	Wisconsin	77.8	100.0	50.0
Missouri	0.0	0.0	0.0	Wyoming	0.0	—	0.0

Source: Calculated by the authors.
a. Dashes indicate no Democrats in delegation.
b. Dashes indicate no Republicans in delegation.
c. No incumbent in First District.

the other hand, are predominantly from the Midwest and South because those two regions would have to achieve larger aggregate reductions in emissions, coupled with some potential losses in coal market shares. This pattern is attributable to the fact that the redistributional effects of acid rain controls are substantial, while consensus on goals remains low. Thus, while H.R. 4567 was approved by a lopsided 16-9 vote on May 20, 1986, by the Subcommittee on Health and Environment, it was not considered for either hearings or markup by the full Committee on Energy and Commerce prior to the end of the Ninety-ninth Congress. This is not surprising, since the full committee was about evenly divided between supporters and opponents of controlling acid rain and its chairman, John Dingell (D-Mich.), has opposed previous attempts to enact such legislation.

The Senate Committee on Environment and Public Works held hearings but failed to debate acid rain legislation even though its chairman, Robert Stafford (R-Vt.), sponsored legislation designed to prevent wholesale fuel switching from high-sulfur to low-sulfur coal. With the Democratic party becoming the majority party in the Senate for the One-hundredth Congress, the committee will be chaired by Quentin Burdick (D-N.D.). Like his predecessor, Senator Burdick is a supporter of acid rain controls and was a cosponsor of S. 2831, which would have permitted switching to reduce SO_2 emissions by approximately 10 million tons by 1997. In addition, a new Subcommittee on Environmental Protection to be chaired by George Mitchell (D-Maine), another supporter of emissions reductions, was created. However, the new Senate majority leader, Robert Byrd (D-W.V.), has consistently opposed legislation which might affect adversely high-sulfur coal production. As a result, while consensus is growing that further emissions reductions are inevitable in the future, similar agreement on the desirability of a cost-effective, flexible law remains elusive.

Acid rain forces policymakers to confront what increasingly has become the norm for environmental controversies. While a scientific foundation for some type of government action exists, popular perceptions of the risk of not acting are apparently more powerful motivations for government intervention than actual scientific evidence (Boyle and Boyle 1983; Howard and Perley 1980). Neither the data nor the scientific knowledge exists to delineate comprehensively and precisely the effects of acid rain or the efficiency of control measures on a site-specific basis (Regens and Donnan 1986). In fact, the knowledge needed to implement an environmental quality-based standard for acid deposition may not be available for at least a decade (U.S. Office of Technology Assessment 1984). As a result, because the emphasis on controls is based on pollution control at emission sources rather than on environmentally efficient goals or standards, the possible costs are very large while the benefits remain conjectural. On the other hand, there is sufficient evidence regarding cause and effect as well as growing political pressure to persuade an increasing number of parties-at-interest that some action to remedy the problem should be taken now. Moreover, the United States has a long tradition of undertaking regulatory action to demonstrate its concern about a problem, or simply to do something. And, as a society, we often implement risk-reduction strategies on the basis of only fragmentary evidence of actual hazards or of the costs and benefits of alternative strategies (Crandall and Lave 1981).

In such circumstances, the economic efficiency of any control option and financial strategy may be of only marginal importance relative to its broader political aspects. Patterns of interest-group mobilization and interaction, including how interest groups use information from the physical and natural sciences or economics, are probably more significant in determining the ultimate resolution of the acid rain debate. In other words, classic distributional issues—who gains the benefits and who bears the burden of the costs and risks—are likely to drive the decision to embark on any particular regulatory path (Rhodes and Middleton 1983). Ultimately, acid rain control decisions are likely to be resolved according to such distributive calculations since they represent an explicit political choice.

At the present time, the environmental and economic issues we have discussed have apparently made it difficult to form a congressional coalition for enacting legislation requiring substantial emissions reductions (Crandall 1984). And, in the absence of new coalitions, the politics of acid rain is likely to continue to be a surrogate for the broader debate over air quality management. Moreover, while the notion of a phased reduction, perhaps through the United States' adoption of the 30 percent proposal, has definite appeal since control costs would be lower and research findings could be incorporated, this proposal does not eliminate all political obstacles. Even if the first stage does not trigger forced scrubbing, it still would affect high-sulfur coal interests.

As a consequence, without dramatic evidence that acid rain is an ecological disaster requiring some type of prompt regulatory intervention, it is unlikely that a coalition of the sort that created the current Clean Air Act can be assembled. Moreover, in the absence of such dramatic findings, pressures will mount to design an environmentally effective or economically efficient strategy proportionate to the magnitude of the risk. Such legislation would represent a nonincremental shift in the nation's approach to air quality management. However, it is difficult to envision when a majority favoring such a policy might emerge without strong presidential support.

Despite these barriers to consensus formation, a phased reduction with a substantial initial target might offer some basis for compromise. For example, setting the reductions to be achieved at the 30 percent level based on annual average estimates for 1980 SO_2 and NO_x emissions might be feasible. That is a sufficiently large enough drop to test the proportionality of the relationship between emissions rollbacks and deposition. At the same time, if coupled with liming on a selective basis, it might prevent or mitigate damages to sensitive resources, especially aquatic ecosystems. Ultimately, such a phased approach, linked to a "polluter-pays" strategy, might be the best hope for compromise. This is not only because of lowered control costs of such an approach and the continuing opportunity for the incremental use of scientific knowledge, but also because of ongoing attempts to coordinate environmental policy with reductions in federal deficits and spending levels.

This is not a happy scenario for environmentalists, but it could be modified. What sort of changes would be necessary to move the policymaking process away from what some (Stanfield 1986a; Freeman 1985) have termed a political stalemate over acid rain? Essentially, it can be argued, significant modifications in any of the four main components of the controversy—science, technology, economics, and politics—might be sufficient to fracture the stalemate. What then are the prospects for changes in each area?

Prospects for Policymaking 159

For national acid rain controls to be implemented in the United States, one or more of the following events must take place. Better scientific understanding of the causes and effects of acid deposition may alter perceptions about the problem as well as appropriate responses to it. A major advancement in technological capabilities to control the precursors of acid rain could alter the feasibility of various policy options, especially retrofitting older facilities not subject to NSPS in order to reduce their emissions. More sophisticated economic analysis of costs and benefits of policy options may alter willingness to undertake various control measures. A reconfiguration of political alliances and coalitions, based upon shifting perceptions of the environmental and economic implications of the acid rain problem, may emerge.

Better Science

It is hard to overemphasize the impact of increased scientific knowledge and how it continues to modify the acid rain debate (see Gould 1985). However, even an aggressive R&D program is of little immediate use; its effects would not be felt for many years. Yet this conventional wisdom could underestimate the potential for more rapid scientific advances.

Two recent studies illustrate this point dramatically. In early 1985, the World Resources Institute, a Washington policy research center, released a report warning of a potential acid rain threat to the western United States. Acknowledging the need for "laying better groundwork for decisions," the report nonetheless concluded that "clear evidence of acidic deposition on sensitive ecosystems in the West and the possibility that chemical and biological damages may be occurring warrant actions in addition to research" (Roth et al. 1985, 33). Six months later a study was released by the Environmental Defense Fund. The EDF analysis, conducted by several of the same scientists who had worked on the World Resources Institute study, claimed to demonstrate a linear relationship between sulfur dioxide emissions from nonferrous metal smelters in the western United States and sulfate concentrations in precipitation in the Rocky Mountain states (Oppenheimer et al. 1985).

Both of the studies have generated tremendous debate within the scientific community. Numerous government and academic analysts have called for caution in interpreting the linearity thesis, given the limited data used in the EDF analysis. EPA representatives have been particularly critical of the studies. For example, Dwaine Winters, the acting director of the agency's acid rain office, asserted, "Our data provide the only direct evidence from exclusively wet deposition monitors on the linearity question for North America." He went on to say, "Based on our review of the work so far, we believe it does not effectively establish a direct source-receptor relationship" (Ember 1985, 19). Nevertheless, if validated by additional data, findings such as these could dramatically modify the acid rain debate. Because the western states have generally opposed any majoritarian solution to financing acid rain controls, a convincing scientific case for the existence of an acid rain problem in their own backyard could persuade western states to form a new alliance of cost-bearers with northeastern states and environmental interests (see Sun 1985).

Technological Breakthrough

Because our approach to environmental mitigation has been based largely on technological fixes, any dramatic innovation on the technical front could alter the existing stalemate and transform the political calculus. While a number of clean coal technologies are being developed, current R&D funding levels appear too limited to expedite progress (Abelson 1985). Nonetheless, more rapid technological developments are not beyond the pale, especially by the mid-1990s.

Some experts believe that the limestone injection multistage burner is the strongest candidate as a technological answer to the acid rain controversy. According to analyses supported by DOE (Energy Research and Advisory Board 1985), LIMB appears to be more promising than flue gas desulfurization on both economic and energy efficiency grounds. Yet other emerging technologies may represent even more significant improvements. In this category are such alternatives as integrated gasification–combined cycle and pressurized fluidized bed combustion systems. However, unlike LIMB, these options are not projected for early commercialization (Catalano and Makansi 1984). Accelerating commercialization for any of these technologies appears to demand much more cooperation between the public and private sectors than has been evidenced to date. The Reagan administration's policy of allowing market forces to govern the demonstration and deployment activities in the energy field potentially represents a major barrier to a public-private partnership (see Lefevre 1985; Parker and Trumbule 1983).

More Comprehensive Economic Analysis

Scientific and technological advances will provide stronger underpinnings for economic analysis of the benefits and costs of acid rain controls in the future. But advances in our economic understanding of these issues are not entirely a function of increased scientific knowledge or engineering capability (see Regens and Donnan 1986; Mandelbaum 1985; Crocker and Regens 1985). More sophisticated economic models of the energy-environment relationship could help overcome the acid rain stalemate. This is particularly true of the benefits of controlling acid rain. As Richard Gordon (1984, 23) puts it: "The vital questions of pollution policy cannot be resolved by available models. We cannot use them to determine the benefits of pollution control. . . . Good models of the impacts of pollution are unavailable. In fact, one is hard put to find clear statements of any kind of just what problems are being cured." As we noted earlier, a clear delineation of the economic benefits of acid rain mitigation seems essential to a resolution of the current impasse.

Political Pragmatism

The acid rain issue has mirrored the larger energy-environment debate in its tendency toward ideological polarization. For any dramatic shift in the status quo, there must be a return to more pragmatic approaches (see Freeman 1985). At present, compromise between the beneficiaries and cost-bearers of acid rain mitigation seems a remote possibility. And yet there are indications that the debate may be shifting slowly toward strategies incorporating more pragmatic options. For example, there is increasing discussion of the merits of more deliberate approaches that permit ongoing adjustments of policy goals and program elements as new information becomes available. These might include various phased-in programs, perhaps augmented by an interim lake-liming effort (Freeman 1985).

Some observers suggest considering hybrid programs, combining selected elements of the various options for funding and control. For example, a national electricity tax might be combined with some assurance of a return in revenue to each state (Sununu 1985). Still others assert that there may be advantages in decentralizing mitigation efforts. This might involve giving the states a greater role in determining their own control schemes (Stanfield 1984). Certainly these do not exhaust the range of possibilities. But institutional innovation in acid rain policy, as with most scientific and technological issues, has not kept pace with knowledge or technology (see Chartock et al. 1985). Developing a consensus on acid rain mitigation, however, appears to demand just such imagination and innovation.

Implications for the Future

The ongoing controversy surrounding acid rain illustrates the highly incremental nature of policymaking for air quality management. Decision-makers risk making costly errors in their policy choices. Regens and Donnan (1986, 343) observe that these errors can take either of two forms:

If SO_2 emissions and associated sulfate ($SO_4^=$) deposition are not a primary cause of adverse effects and a major commitment to SO_2 abatement is made, substantial resources will be taken from other beneficial uses without achieving the desired results.

If SO_2 emissions and $SO_4^=$ deposition are the primary cause of widespread adverse effects and abatement programs are deferred, then an irreversible degree of environmental damage may possibly occur.

Figure 13 depicts this policy dilemma in simplified terms. Basically, the choice is to act now, accepting the underlying uncertainty in existing scientific information, or to wait until more is known, being conscious of the social penalties of error implicit in each of the two choices.

Moreover, at any given time, finite public and private resources must be divided among many competing demands. When either governments or markets channel resources to meet some specified demand such as acid rain control, other public and/or private objectives are subject to an inescapable

Prospects for Policymaking 163

**Figure 13
The Policy Dilemma**

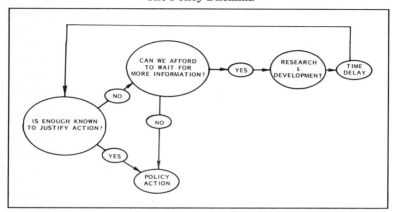

tradeoff (see Rhoads 1985). Given this obvious constraint, it is apparent that in a democratic polity it is important to identify what must be foregone to achieve various policy goals as well as who will be affected, for better and for worse, under different options. As a result, the key question facing policymakers involves a judgment as to whether further action to reduce emissions is justifiable and necessary in view of the potential benefits and costs of such action.

While science has already identified problems caused by acid rain, and possible options to prevent or mitigate its adverse effects, additional research is needed to determine the extent and rate of damages, especially to nonaquatic resources, on a nationwide basis. Moreover, the absence of a clear and immediate public health threat has forestalled action in the face of uncertainty. What is more readily understood is the high cost of control, particularly when coupled with imprecise estimates of benefits. This has fostered opposition by potential cost-bearing interests. Yet if benefits are viewed as not just the avoidance of damages to nature but also include improved visibility and the salutary effects of cleaner and purer air, then the relationship between costs and benefits becomes more favorable.

Because of the growing evidence that acid rain is a truly

national problem, even though its impact varies from region to region, a market-oriented policy that incorporates the "polluter-pays" financing option emerges as equitable as well as defensible. In fact, even if regional expenditures for pollution control were fixed at 1979 levels, a cost-minimizing allocation of emissions reduction requirements would result in an additional drop in SO_2 emissions ranging from 1.3 percent in the Midwest to 33.2 percent in the Northeast, 11.1 percent in the Great Lakes region, 24.4 percent in the South, and 28.6 percent in the West (Gollop and Roberts 1985). The important implication for policymaking is that it is possible to allocate emissions among utilities in a manner that either reduces the overall costs of achieving the current level of emissions within each region or achieves a lower level of emissions at the current overall cost. Such a strategy allows one to phase in reductions in order to minimize national and regional economic dislocations. It also reduces the likelihood that acid rain will pose a serious environmental threat in areas where it is not yet a major problem. Finally, it goes beyond a merely symbolic call for action. Instead, it offers a framework for establishing a compromise between contending interests that reconciles the competing desires for economic efficiency, environmental effectiveness, and social equity.

There is mounting evidence of a causal relationship between sulfur emissions produced by burning fossil fuels and acid rain (National Research Council 1986). More and more people are convinced that we now know enough to start solving the problem by restricting sulfur and nitrogen oxides emissions. However, research since 1972 provides only limited evidence of substantial nonaquatic damages attributable to acid rain (see Katzenstein 1986). The lack of widespread, quantifiable effects coupled with the high cost of control discourages but does not preclude prompt action to deal with the acid rain issue in an effective and balanced way. Nonetheless, bearing in mind such major obstacles to achieving consensus about an appropriate policy response, government action is inevitable because the acid rain issue has increasingly come to be perceived by both political leaders and the general public as a serious environmental and transboundary problem. James L. Regens, then EPA joint chairman of the Interagency Task Force on Acid Precipitation, noted in 1982: "The real policy question is not whether regulatory actions will be taken but when such measures will commence.... In the interim we should do nothing that might exacerbate the problem" (quoted in Mosher 1982, 457). As a result, those who shape policy should reach decisions with their eyes open to the implications of their choices.

The acid rain issue is not a unique case. The problems of ambiguities in science, disparities in the distribution of benefits and costs, and acrimonious politics are increasingly common in contemporary environmental issues. Total understanding of the problem will never be achieved. Bringing existing scientific and economic information to bear permits a rational determination of the parameters of the problem; it also raises the possibility of identifying the technically viable options for addressing it. Regulatory action can be taken to the extent that any of those options are politically feasible. Real hopes for the adoption of a national policy that goes beyond "we-must-wait-and-study-more," therefore, rests primarily on the emergence of a political consensus to act. Enhanced scientific knowledge or better understanding of the benefits and costs can help to clarify the potential consequences of such action, but they are not sufficient substitutes for the political will to make a choice.

APPENDICES

NOTES

REFERENCES

INDEX

APPENDIX A

Top 100 SO$_2$ Emitting Plants

Utility/Plant	Unit Number	Year of Service	Generation Capacity (net megawatts)	Emissions[a] SO$_2$	NO$_x$
Alabama Power/Gaston	1	1960	272	26.1	8.8
	2	1960	272	25.5	8.7
	3	1961	272	24.2	8.2
	4	1962	245	23.2	7.8
	5	1974	952	70.1	16.6
Alabama Power/Gorgas	5	1944	69	3.1	1.1
	6	1951	125	5.1	2.4
	7	1952	125	7.0	3.3
	8	1956	188	12.3	4.2
	9	1958	190	13.8	4.8
	10	1972	789	43.0	14.8
Appalachian Power/Amos	1	1971	816	33.3	23.5
	2	1972	816	29.9	21.1
	3	1973	1,300	41.9	29.6
Associated Electric Coop/New Madrid	1	1972	600	198.9	34.2
Associated Electric Coop/Thomas Hill	1	1966	180	31.0	9.0
	2	1969	290	63.9	18.7
Big Rivers Electric/Coleman	1	1969	166	9.1	7.9
	2	1970	166	27.6	8.5
	3	1972	170	31.6	9.2
Big Rivers Electric/Henderson II	1	1973	175	25.3	6.2
	2	1974	175	24.9	6.1
Buckeye Power/Cardinal	1	1967	615	40.0	12.0
	2	1967	615	37.3	11.2
	3	1977	650	48.7	14.6
Carolina Power and Light/Roxboro	1	1966	411	14.9	12.8
	2	1968	657	26.6	16.4
	3	1973	745	28.9	17.8
Central Illinois Public Service/Coffeen	1	1965	389	53.1	31.5
	2	1972	616	84.2	49.8
Central Illinois Public Service/Meredosia	1	1948	58	17.1	0.9
	2	1948	58	16.0	0.8

Utility/Plant	Unit Number	Year of Service	Generation Capacity (net megawatts)	Emissions[a]	
				SO_2	NO_x
	3	1949	58	15.0	0.8
Cincinnati Gas and Electric/Beckjord	1	1952	100	7.2	2.8
	2	1953	100	7.7	2.9
	3	1954	125	9.7	5.2
	4	1958	163	10.2	3.9
	5	1962	240	21.7	8.3
	6	1969	434	27.9	10.5
Cincinnati Gas and Electric/Miami Fort	5	1949	100	4.7	1.4
	6	1960	168	13.6	2.8
	7	1975	512	35.9	10.4
	8	1978	512	35.3	10.2
City of Springfield, Illinois/Dallman	1	1968	80	9.0	3.2
	2	1972	80	10.0	3.6
	3	1978	192	26.6	3.0
Cleveland Electric Illuminating/ Ashtabula	5	1948	46	6.6	1.1
	6	1948	46	6.6	1.1
	7	1958	256	41.7	6.9
	8	1948	46	6.6	1.1
	9	1948	46	6.9	1.1
Cleveland Electric Illuminating/ Eastlake	1	1953	123	14.5	2.1
	2	1953	123	15.1	2.2
	3	1954	123	16.6	2.4
	4	1956	208	20.4	2.9
	5	1972	680	88.4	17.7
Cleveland Electric Illuminating/Avon Lake	6	1949	86	7.1	1.5
	7	1949	86	7.1	1.5
	8	1959	233	19.2	4.2
	9	1970	680	56.1	12.2
Columbus and Southern Ohio/ Conesville	1	1959	115	19.6	5.8
	2	1957	115	21.8	6.5
	3	1962	161	28.5	3.8
	4	1973	800	118.7	11.4
	5	1976	375	5.8	5.6
	6	1978	375	6.4	6.1
Commonwealth Edison/Kincaid	1	1967	660	84.8	13.6
	2	1968	660	120.7	19.7
Consumers Power/Campbell	1	1962	265	3.2	13.2
	2	1967	385	20.9	4.9
	3	1980	778	30.9	7.2
Consumers Power/Cobb	1	1948	66	4.3	0.9
	2	1948	66	4.6	1.0
	3	1950	66	5.0	1.0
	4	1956	156	17.9	3.7
	5	1957	156	16.7	3.5
Dayton Power and Light/Stuart	1	1971	610	28.7	14.6
	2	1970	610	27.6	13.9
	3	1972	610	28.8	14.6
	4	1974	610	27.7	13.9

Utility/Plant	Unit Number	Year of Service	Generation Capacity (net megawatts)	Emission[a] SO$_2$	NO$_x$
Detroit Edison/Monroe	1	1971	817	70.4	20.0
	2	1973	823	46.7	13.3
	3	1973	823	59.4	16.9
	4	1974	817	57.4	16.3
Detroit Edison/St. Clair	1	1953	169	4.9	4.6
	2	1953	156	3.9	3.6
	3	1954	156	4.6	4.2
	4	1954	169	5.2	4.9
	6	1961	353	7.9	5.3
	7	1969	544	17.9	11.9
Duke Power/Belews Creek	1	1974	1,080	45.5	26.4
	2	1975	1,080	41.1	23.9
Duke Power/Marshall	1	1965	350	15.7	8.4
	2	1966	350	14.7	7.9
	3	1969	650	26.5	14.2
	4	1970	650	28.1	15.1
Electric Energy/Joppa Steam	1	1953	181	17.4	4.7
	2	1953	181	16.9	4.6
	3	1954	181	16.4	4.4
	4	1954	181	14.8	4.0
	5	1955	181	17.2	4.7
	6	1955	181	17.3	4.7
Empire Distric Electric/Asbury	1	1970	200	65.0	9.0
Florida Power/Crystal River	1	1966	440	33.7	19.8
	2	1969	524	38.9	22.8
Georgia Power/Bowen	1	1971	700	40.5	12.7
	2	1972	700	59.3	18.1
	3	1974	880	74.2	22.6
	4	1975	880	74.1	22.6
Georgia Power/Hammond	1	1954	125	5.6	1.8
	2	1954	125	6.1	2.0
	3	1955	125	6.3	2.0
	4	1970	578	29.3	9.5
Georgia Power/Harllee Branch	1	1965	250	11.9	5.3
	2	1967	319	12.9	5.8
	3	1968	481	14.2	6.3
	4	1969	491	16.1	7.1
Georgia Power/Harrison	1	1972	684	67.2	13.2
Georgia Power/Wansley	1	1976	952	100.0	25.3
	2	1978	952	109.8	27.9
Georgia Power/Yates	1	1950	100	8.1	2.3
	2	1950	100	8.2	2.3
	3	1953	100	8.9	2.5
	4	1957	125	11.7	3.4
	5	1958	125	11.4	3.3
	6	1974	350	29.6	8.5
	7	1974	350	28.7	8.2
Gulf Power/Crist	4	1959	94	8.5	1.5

Utility/Plant	Unit Number	Year of Service	Generation Capacity (net megawatts)	Emission[a] SO_2	NO_x
	5	1961	94	8.3	1.5
	6	1970	370	12.8	3.2
	7	1973	578	46.7	11.6
Hoosier Division of Indiana Power/ Ratts	1	1969	117	25.4	6.0
	2	1969	117	24.3	5.8
Illinois Power/Baldwin	1	1970	623	86.1	36.8
	2	1973	634	87.9	37.4
	3	1975	634	90.1	12.2
Indiana and Michigan Electric/Breed	1	1960	496	71.0	13.0
Indiana and Michigan Electric/ Tanner's Creek	1	1951	152	10.3	1.8
	2	1952	152	14.3	2.5
	3	1954	215	13.4	2.3
	4	1964	580	56.8	21.8
Indiana Kentucky Electric/Clifty Creek	1	1955	217	58.0	14.0
	2	1955	217	42.6	10.3
	3	1955	217	43.8	10.6
	4	1955	217	46.2	11.2
	5	1955	217	49.7	12.0
	6	1956	217	47.9	11.6
Indianapolis Power and Light/ Petersburg	1	1967	253	27.7	9.2
	2	1969	471	30.5	10.1
	3	1977	531	50.3	16.6
Indianapolis Power and Light/Stout	5	1958	114	23.1	6.9
	6	1961	114	27.0	8.2
Iowa Public Service/Neal (Sargent Bluffs)	1	1964	139	1.8	1.6
	2	1972	349	10.7	9.5
	3	1975	520	13.5	12.0
	4	1979	576	18.3	16.2
Kansas City Power and Light/ Hawthorn	1	1951	69	4.7	1.0
	2	1951	69	3.5	0.8
	3	1953	112	2.8	0.6
	4	1955	143	4.8	1.0
	5	1969	515	29.5	6.4
Kansas City Power and Light/ Montrose	1	1958	188	47.6	6.8
	2	1958	188	58.1	8.4
	3	1958	188	56.3	8.1
Kentucky Power/Big Sandy	1	1963	280	20.6	9.5
	2	1969	816	37.3	17.1
Kentucky Utilities Power/Brown	1	1957	100	6.9	1.6
	2	1963	156	13.0	2.2
	3	1971	438	33.3	5.5
Kentucky Utilities Power/Ghent	1	1973	511	44.6	12.3
	2	1977	511	40.8	11.3
Louisville Gas and Electric/Mill Creek	1	1972	321	44.1	5.7
	2	1974	321	44.1	5.7
	3	1978	411	5.8	10.5
Mississippi Power/Watson	4	1968	299	26.5	26.5
	5	1973	578	39.4	39.4

Appendices 173

Utility/Plant	Unit Number	Year of Service	Generation Capacity (net megawatts)	Emission[a] SO_2	NO_x
Monongahela Power/Fort Martin	1	1967	576	55.5	15.8
	2	1968	576	39.6	11.3
Monongahela Power/Harrison	2	1973	684	74.6	14.7
	3	1974	684	73.2	14.4
New England Power/Brayton Point	1	1963	241	20.4	2.6
	2	1964	241	22.7	3.0
	3	1969	643	40.1	7.4
Niagara Mohawk Power/Dunkirk	1	1950	96	9.5	9.5
	2	1950	96	9.2	9.3
	3	1959	218	13.0	13.0
	4	1960	218	20.3	20.2
Niagara Mohawk Power/Huntley	63	1942	100	5.0	2.3
	64	1948	100	6.4	2.9
	65	1953	100	6.5	3.0
	66	1954	100	6.2	2.8
	67	1957	218	14.6	9.4
	68	1958	218	11.9	7.7
Northern Indiana Public Service/Bailly	7	1962	194	22.1	8.1
	8	1968	422	26.8	6.5
Northern Indiana Public Service/ Michigan City	1	1950	70	8.0	2.9
	2	1951	70	8.0	2.9
	12	1974	521	59.4	21.5
Ohio Edison/Burger	1	1944	31	13.5	0.4
	2	1944	31	17.3	0.5
	3	1947	31	13.1	0.4
	4	1947	31	13.1	0.4
	5	1950	50	27.7	0.8
Ohio Edison/Sammis	1	1959	185	9.2	1.1
	2	1960	185	8.1	1.0
	3	1961	185	4.6	0.6
	4	1962	185	10.2	1.2
	5	1967	318	24.7	3.0
	6	1969	623	42.1	5.1
	7	1971	623	38.1	4.7
Ohio Power/Mitchell	1	1971	816	78.8	16.9
	2	1971	816	92.5	19.9
Ohio Power/Muskingum River	1	1953	220	31.6	4.1
	2	1954	220	34.1	4.4
	3	1957	238	36.5	10.5
	4	1958	238	27.0	7.8
	5	1968	615	114.8	15.0
Ohio Power/Gavin	1	1974	1,300	175.1	38.0
	2	1975	1,300	201.3	43.6
Ohio Power/Kammer	1	1958	238	43.2	9.9
	2	1958	238	50.4	11.6
	3	1959	238	55.6	12.8
Ohio Valley Electric/Kyger Creek	1	1955	217	41.4	9.5
	2	1955	217	42.0	9.6

Utility/Plant	Unit Number	Year of Service	Generation Capacity (net megawatts)	Emission[a] SO$_2$	NO$_x$
	3	1955	217	34.4	7.9
	4	1955	217	42.5	9.7
	5	1955	217	42.0	9.6
Pennsylvania Electric/Conemaugh	1	1970	936	95.6	21.4
	2	1971	936	109.6	24.7
Pennsylvania Electric/Homer City	1	1969	600	54.1	13.5
	2	1969	600	43.8	10.9
	3	1977	650	71.2	17.7
Pennsylvania Electric/Keystone	1	1967	936	61.4	22.3
	2	1968	936	80.9	22.8
Pennsylvania Electric/Shawville	1	1954	132	10.7	3.4
	2	1954	132	14.6	4.7
	3	1959	188	20.1	4.6
	4	1960	188	19.4	4.5
Pennsylvania Power/Mansfield	1	1976	835	19.1	13.4
	2	1977	835	18.4	12.9
	3	1980	835	19.1	13.4
Pennsylvania Power and Light/ Brunner Island	1	1961	363	28.4	7.8
	2	1965	405	39.4	10.9
	3	1969	790	71.2	19.6
Pennsylvania Power and Light/Martins Creek	1	1954	156	32.2	14.4
	2	1956	156	28.1	12.6
Pennsylvania Power and Light/ Montour	1	1971	734	92.9	18.9
	2	1973	800	16.6	20.9
Potomac Electric Power/Chalk Point	1	1964	364	19.7	7.9
	2	1965	364	30.7	12.4
Potomac Electric Power/Morgantown	1	1970	573	39.5	8.1
	2	1971	575	52.5	10.9
Public Service of Indiana/Cayuga	1	1970	531	56.0	13.7
	2	1972	531	59.9	14.7
Public Service of Indiana/Gallagher	1	1959	150	10.1	1.9
	2	1958	150	16.1	3.1
	3	1960	150	18.4	3.5
	4	1961	150	14.9	2.8
Public Service of Indiana/Gibson	1	1976	668	70.6	11.6
	2	1975	668	68.4	11.2
	3	1978	668	82.6	13.5
	4	1979	668	84.0	13.7
Public Service of Indiana/Wabash River	1	1953	112	8.1	1.6
	2	1953	112	8.1	1.6
	3	1954	112	10.3	2.9
	4	1955	113	8.5	1.7
	5	1956	125	7.5	1.5
	6	1968	387	23.4	6.6
Public Service of New Hampshire/ Merrimack	1	1960	114	15.1	5.8
	2	1968	346	37.2	14.2
Southern Indiana G & E/Culley	1	1955	46	6.3	1.0
	2	1966	104	16.9	2.8

Appendices 175

Utility/Plant	Unit Number	Year of Service	Generation Capacity (net megawatts)	Emission[a] SO$_2$	NO$_x$
	3	1973	265	40.5	6.7
Tampa Electric/Big Bend	1	1970	446	59.3	17.5
	2	1973	446	42.8	10.5
	3	1976	446	51.1	15.
Tennessee Valley Authority/Allen	1	1958	330	26.7	11.1
	2	1959	330	23.1	9.3
	3	1959	330	24.1	9.7
Tennessee Valley Authority/Colbert	1	1954	200	21.9	5.3
	2	1955	200	19.9	4.8
	3	1955	223	23.1	5.6
	4	1955	223	26.8	6.5
	5	1962	550	34.2	8.3
Tennessee Valley Authority/ Cumberland	1	1972	1,300	179.7	26.7
	2	1973	1,300	177.1	26.3
Tennessee Valley Authority/Gallatin	1	1956	300	40.9	7.9
	2	1957	300	38.1	7.4
	3	1959	328	25.5	4.9
	4	1959	328	32.6	6.3
Tennessee Valley Authority/ Johnsonville	1	1951	125	9.4	2.6
	2	1951	125	13.5	3.7
	3	1952	125	14.2	3.9
	4	1952	125	15.9	4.4
	5	1952	147	17.4	4.7
	6	1953	147	13.6	3.7
	7	1958	173	11.7	2.3
	8	1959	173	17.1	3.3
	9	1959	173	4.0	0.7
	10	1959	173	18.5	3.6
Tennessee Valley Authority/Kingston	1	1954	175	8.4	4.5
	2	1954	175	7.2	3.8
	3	1954	175	8.0	4.2
	4	1954	175	6.8	3.6
	5	1954	200	12.9	6.8
	6	1955	200	8.8	4.6
	7	1955	200	10.6	5.6
	8	1955	200	13.2	7.0
	9	1955	200	12.7	6.8
Tennessee Valley Authority/Paradise	1	1963	704	132.0	32.9
	2	1963	704	92.1	23.1
	3	1969	1,150	118.6	29.7
Tennessee Valley Authority/Sevier	1	1955	223	19.3	5.2
	2	1955	223	21.9	5.9
	3	1956	200	21.4	5.8
	4	1957	200	18.7	5.0
Tennessee Valley Authority/Shawnee	1	1953	175	8.3	3.2
	2	1953	175	9.5	3.7
	3	1953	175	10.7	4.1
	4	1954	175	10.2	4.0

Utility/Plant	Unit Number	Year of Service	Generation Capacity (net megawatts)	Emission[a] SO$_2$	Emission[a] NO$_x$
	5	1954	175	8.7	3.3
	6	1954	175	9.1	3.6
	7	1954	175	10.3	4.0
	8	1955	175	8.8	3.4
	9	1955	175	9.3	3.6
	10	1956	175	11.6	4.5
Tennessee Valley Authority/Widows Creek	1	1952	141	5.9	4.3
	2	1952	141	6.2	4.5
	3	1952	150	5.7	3.9
	4	1953	141	4.0	2.9
	5	1954	141	6.5	4.7
	6	1954	141	6.5	4.7
	7	1960	575	31.5	32.0
	8	1964	550	1.0	10.5
Union Electric/Labadie	1	1970	555	62.5	14.2
	2	1971	555	44.3	10.0
	3	1972	555	63.9	14.5
	4	1973	555	66.5	15.0
Union Electric/Rush Island	1	1976	571	26.7	15.4
	2	1977	571	25.2	14.5
Union Electric/Sioux	1	1967	550	49.5	19.7
	2	1967	550	47.7	18.9
Virginia Electric and Power/Mt. Storm	1	1965	570	26.0	7.8
	2	1966	570	31.5	9.4
	3	1973	522	34.1	10.2
West Penn Power/Hatfield's Ferry	1	1969	576	46.2	10.0
	2	1970	576	60.3	13.0
	3	1971	576	64.7	13.9
Wisconsin Electric Power/N. Oak Creek	1	1953	120	13.3	3.2
	2	1954	120	7.5	1.2
	3	1955	130	11.4	2.7
	4	1957	130	13.7	3.3
Wisconsin Electric Power/Port Washington	1	1935	80	7.6	0.0
	2	1943	80	10.4	0.0
	3	1948	80	10.1	0.0
	4	1949	80	7.9	0.0
	5	1950	80	10.0	0.0
Wisconsin Electric Power/S. Oak Creek	5	1959	275	16.8	3.5
	6	1961	275	18.2	3.8
	7	1965	310	18.4	3.8
	8	1967	310	24.3	5.1
Wisconsin Power and Light/Edgewater	1	1931	30	0.8	0.1
	2	1931	30	0.6	0.1
	3	1951	60	5.6	1.9
	4	1969	330	51.0	17.2

Source: Adapted from Hall (1985).

a. Estimates in thousands of tons per year for 1980 using unit MW and capacity factors for each boiler. Firing type—cyclone, wall, wall/opposed, tangential—also used for NO$_x$ estimates.

APPENDIX B

Convention on Long-Range Transboundary Air Pollution

The Parties to the Present Convention,

Determined to promote relations and co-operation in the field of environmental protection,

Aware of the significance of the activities of the United Nations Economic Commission for Europe in strengthening such relations and co-operation, particularly in the field of air pollution including long-range transport of air pollutants,

Recognizing the contribution of the Economic Commission for Europe to the multilateral implementation of the pertinent provisions of the Final Act of the Conference on Security and Co-operation in Europe,

Cognizant of the references in the chapter on environment of the Final Act of the Conference on Security and Co-operation in Europe calling for co-operation to control air pollution and its effects, including long-range transport of air pollutants, and to the development through international co-operation of an extensive programme for the monitoring and evaluation of long-range transport of air pollutants, starting with sulphur dioxide and with possible extension to other pollutants,

Considering the pertinent provisions of the Declaration of the United Nations Conference on the Human Environment, and in particular principle 21, which expresses the common conviction that States have, in accordance with the Charter of the United Nations and the principles of international law, the sovereign right to exploit their own resources pursuant to their own environmental policies, and the responsibility to ensure that activities within their jurisdiction or control do not cause damage to the environment of other States or of areas beyond the limits of national jurisdiction,

Recognizing the existence of possible adverse effects, in the short and long term, of air pollution including transboundary air pollution,

Concerned that a rise in the level of emissions of air pollutants within the region as forecast may increase such adverse effects,

Recognizing the need to study the implications of the long-range transport of air pollutants and the need to seek solutions for the problems identified,

Affirming their willingness to reinforce active international cooperation to develop appropriate national policies and by means of exchange of information, consultation, research and monitoring, to coordinate national action for combatting air pollution including long-range transboundary air pollution,

Have agreed as follows:

Definitions

Article 1

For the purposes of the present Convention:

(a) *"air pollution"* means the introduction by man, directly or indirectly, of substances or energy into the air resulting in deleterious effects of such a nature as to endanger human health, harm living resources and ecosystems and material property and impair or interfere with amenities and other legitimate uses of the environment, and "air pollutants" shall be construed accordingly;

(b) *"long-range transboundary air pollution"* means air pollution whose physical origin is situated wholly or in part within the area under the national jurisdiction of one State and which has adverse effects in the area under the jurisdiction of another State at such a distance that it is not generally possible to distinguish the contribution of individual emission sources or groups of sources.

Fundamental Principles

Article 2

The Contracting Parties, taking due account of the facts and problems involved, are determined to protect man and his environment against air pollution and shall endeavour to limit and, as far as possible, gradually reduce and prevent air pollution including long-range transboundary air pollution.

Article 3

The Contracting Parties, within the framework of the present Convention, shall by means of exchanges of information, consultation, research and monitoring, develop without undue delay policies and strategies which shall serve as a means of combatting the discharge of air pollutants, taking into account efforts already made at national and international levels.

Article 4

The Contracting Parties shall exchange information on and review their policies, scientific activities and technical measures aimed at combatting, as far as possible, the discharge of air pollutants which may have adverse effects, thereby contributing to the reduction of air pollution including long-range transboundary air pollution.

Article 5

Consultations shall be held, upon request, at an early stage between, on the one hand, Contracting Parties which are actually affected by or exposed to a significant risk of long-range transboundary air pollution and, on the other hand, Contracting Parties within which and subject to whose jurisdiction a significant contribution to long-range transboundary air pollution originates, or could originate, in connection with activities carried on or contemplated therein.

Air Quality Management

Article 6

Taking into account articles 2 to 5, the ongoing research, exchange of information and monitoring and the results thereof, the cost and effectiveness of local and other remedies and, in order to combat air pollution, in particular that originating from new or rebuilt installations, each Contracting Party undertakes to develop the best policies and strategies including air quality management systems and, as part of them, control measures compatible with balanced development, in particular by using the best available technology which is economically feasible and low- and non-waste technology.

Research and Development

Article 7

The Contracting Parties, as appropriate to their needs, shall initiate and cooperate in the conduct of research into and/or development of:

(a) existing and proposed technologies for reducing emissions of sulphur compounds and other major air pollutants, including technical and economic feasibility, and environmental consequences;

(b) instrumentation and other techniques for monitoring and measuring emission rates and ambient concentrations of air pollutants;

(c) improved models for a better understanding of the transmission of long-range transboundary air pollutants;

(d) the effects of sulphur compound and other major air pollutants on human health and the environment, including agriculture, forestry, materials, aquatic and other natural ecosystems and visibility, with a view to establishing a scientific basis for dose/effect relationships designed to protect the environment;

(e) the economic, social and environmental assessment of alternative measures for attaining environmental objectives including the reduction of long-range transboundary air pollution;

(f) education and training programs related to the environmental aspects of pollution by sulphur compounds and other major air pollutants.

Exchange of Information

Article 8

The Contracting Parties, within the framework of the Executive Body referred to in article 10 and bilaterally, shall, in their common interests, exchange available information on:

(a) data on emissions at periods of time to be agreed upon, of agreed air pollutants, starting with sulphur dioxide, coming from grid-units of agreed size; or on the fluxes of agreed air pollutants, starting with sulphur dioxide, across national borders, at distances and at periods of time to be agreed upon;

(b) major changes in national policies and in general industrial development, and their potential impact, which would be likely to cause significant changes in long-range transboundary air pollution;

(c) control technologies for reducing air pollution relevant to long-range transboundary air pollution;

(d) the projected cost of the emission control of sulphur compounds and other major air pollutants on a national scale;

(e) meteorological and physico-chemical data relating to the processes during transmission;

(f) long-range transboundary air pollution and the extent of damage[1] which these data indicate can be attributed to long-range transboundary air pollution;

(g) national, subregional and regional policies and strategies for the control of sulphur compounds and other major air pollutants.

---

[1] The present Convention does not contain a rule on State liability as to damage.

Implementation and Further Development of the Co-operative Programme for the Monitoring and Evaluation of the Long-Range Transmission of Air Pollutants in Europe

Article 9

The Contracting Parties stress the need for the implementation of the existing "Co-operative programme for the monitoring and evaluation of the long-range transmission of air pollutants in Europe" (hereinafter referred to as EMEP) and, with regard to the further development of this program, agree to emphasize:

(a) the desirability of Contracting Parties joining in and fully implementing EMEP which, as a first step, is based on the monitoring of sulphur dioxide and related substances;

(b) the need to use comparable or standardized procedures for monitoring whenever possible;

(c) the desirability of basing the monitoring programme on the framework of both national and international programmes. The establishment of monitoring stations and the collection of data shall be carried out under the national jurisdiction of the country in which the monitoring stations are located;

(d) the desirability of establishing a framework for a cooperative environmental monitoring programme, based on and taking into account present and future national, subregional, regional and other international programmes;

(e) the need to exchange data on emissions at periods of time to be agreed upon, of agreed air pollutants, starting with sulphur dioxide, coming from grid-units of agreed size; or on the fluxes of agreed air pollutants, starting with sulphur dioxide, across national borders, at distances and at periods of time to be agreed upon. The method, including the model, used to determine the fluxes, as well as the method, including the model, used to determine the transmission of air pollutants based on the emissions per grid-unit, shall be made available and periodically reviewed, in order to improve the methods and the models;

(f) their willingness to continue the exchange and periodic updating of national data on total emissions of agreed air pollutants, starting with sulphur dioxide;

(g) the need to provide meteorological and physico-chemical data relating to processes during transmission;

(h) the need to monitor chemical components in other media such as water, soil and vegetation, as well as a similar monitoring programme to record effects on health and environment;

Executive Body

Article 10

1. The representatives of the Contracting Parties shall, within the framework of the Senior Advisors to ECE Governments on Environmental Problems, constitute the Executive Body of the present Convention, and shall meet at least annually in that capacity.
2. The Executive Body shall:
    (a) review the implementation of the present Convention;
    (b) establish, as appropriate, working groups to consider matters related to the implementation and development of the present Convention and to this end to prepare appropriate studies and other documentation and to submit recommendations to be considered by the Executive Body;
    (c) fulfill such other functions as may be appropriate under the provisions of the present Convention.
3. The Executive Body shall utilize the Steering Body for the EMEP to play an integral part in the operation of the present Convention, in particular with regard to data collection and scientific co-operation.
4. The Executive Body, in discharging its functions, shall, when it deems appropriate, also make use of information from other relevant international organizations.

Article 11

The Executive Secretary of the Economic Commission for Europe shall carry out, for the Executive Body, the following secretariat functions:

(a) to convene and prepare the meetings of the Executive Body;

(b) to transmit to the Contracting Parties reports and other information received in accordance with the provisions of the present Convention;

(c) to discharge the functions assigned by the Executive Body.

Amendments to the Convention

Article 12

1. Any Contracting Party may propose amendments to the present Convention.
2. The text of proposed amendments shall be submitted in writing to the Executive Secretary of the Economic Commission for Europe, who

shall communicate them to all Contracting Parties. The Executive Body shall discuss proposed amendments at its next annual meeting provided that such proposals have been circulated by the Executive Secretary of the Economic Commission for Europe to the Contracting Parties at least ninety days in advance.

3. An amendment to the present Convention shall be adopted by consensus of the representatives of the Contracting Parties, and shall enter into force for the Contracting Parties which have accepted it on the ninetieth day after the date on which two-thirds of the Contracting Parties have deposited instruments of acceptance with the depositary. Thereafter, the amendment shall enter into force for any other Contracting Party on the ninetieth day after the date on which that Contracting Party deposits its instrument of acceptance of the amendment.

Settlement of Disputes

Article 13

If a dispute arises between two or more Contracting Parties to the present Convention as to the interpretation or application of the Convention, they shall seek a solution by negotiation or by any other method of dispute settlement acceptable to the parties to the dispute.

Signature

Article 14

1. The present Convention shall be open for signature at the United Nations Office at Geneva from 13 to 16 November 1979 on the occasion of the High-Level Meeting within the framework of the Economic Commission for Europe on the Protection of the Environment, by the member States of the Economic Commission for Europe as well as States having consultative status with the Economic Commission for Europe, pursuant to paragraph 8 of Economic and Social Council resolution 36 (IV) of 28 March 1947, and by regional economic integration organizations, constituted by sovereign States members of the Economic Commission for Europe, which have competence in respect of the negotiation, conclusion and application of international agreements in matters covered by the present Convention.

2. In matters within their competence, such regional economic integration organizations shall, on their own behalf, exercise the rights and fulfill the responsibilities which the present Convention attributes to their member States. In such cases, the member States of these organizations shall not be entitled to exercise such rights individually.

Ratification, Acceptance, Approval and Accession

Article 15

1. The present Convention shall be subject to ratification, acceptance or approval.
2. The present Convention shall be open for accession as from 17 November 1979 by the States and organizations referred to in article 14, paragraph 1.
3. The instruments of ratification, acceptance, approval or accession shall be deposited with the Secretary-General of the United Nations, who will perform the functions of the depositary.

Entry into Force

Article 16

1. The present Convention shall enter into force on the ninetieth day after the date of deposit of the twenty-fourth instrument of ratification, acceptance, approval or accession.
2. For each Contracting Party which ratifies, accepts or approves the present Convention or accedes thereto after the deposition of the twenty-fourth instrument of ratification, acceptance, approval or accession, the Convention shall enter into force on the ninetieth day after the date of deposit by such Contracting Party of its instrument of ratification, acceptance, approval or accession.

Withdrawal

Article 17

At any time after five years from the date on which the present Convention has come into force with respect to a Contracting Party, that Contracting Party may withdraw from the Convention by giving written notification to the depositary. Any such withdrawal shall take effect on the ninetieth day after the date of its receipt by the depositary.

Authentic Texts

Article 18

The original of the present Convention, of which the English, French and Russian texts are equally authentic, shall be deposited with the Secretary-General of the United Nations.

WITNESS WHEREOF the undersigned, being duly authorized thereto, have signed the present Convention.

DONE at Geneva, this thirtieth day of November, one thousand nine hundred and seventy-nine.

APPENDIX C

Memorandum of Intent
Between the Government of Canada and
the Government of the United States of America
Concerning Transboundary Air Pollution

The Government of Canada and the Government of the United States of America,

Share a concern about actual and potential damage resulting from transboundary air pollution, (which is the short and long range transport of air pollutants between their countries), including the already serious problem of acid rain;

Recognize this is an important and urgent bilateral problem as it involves the flow of air pollutants in both directions across the international boundary, especially the long range transport of air pollutants;

Share also a common determination to combat transboundary air pollution in keeping with their existing international rights, obligations, commitments and cooperative practices, including those set forth in the 1909 Boundary Waters Treaty, the 1972 Stockholm Declaration on the Human Environment, the 1978 Great Lakes Water Quality Agreement, and the 1979 ECE Convention on Long Range Transboundary Air Pollution;

Undertook in July 1979 to develop a bilateral cooperative agreement on air quality which would deal effectively with transboundary air pollution;

Are resolved as a matter of priority both to improve scientific understanding of the long range transport of air pollutants and its effects and to develop and implement policies, practices and technologies to combat its impact;

Are resolved to protect the environment in harmony with measures to meet energy needs and other national objectives;

Note scientific findings which indicate that continued pollutant loadings will result in extensive acidification in geologically sensitive areas during the coming years, and that increased pollutant loadings will accelerate this process;

Are concerned that environmental stress could be increased if action is not taken to reduce transboundary air pollution;

Are convinced that the best means to protect the environment from the effects of transboundary air pollution is through the achievement of necessary reductions in pollutant loadings;

Are convinced also that this common problem requires cooperative action by both countries;

Intend to increase bilateral co-operative action to deal effectively with transboundary air pollution, including acid rain.

In particular, the Government of Canada and the Government of the United States of America intend:

1. to develop a bilateral agreement which will reflect and further the development of effective domestic control programs and other measures to combat transboundary air pollution;
2. to facilitate the conclusion of such an agreement as soon as possible; and,
3. pending conclusion of such an agreement, to take interim actions available under current authority to combat transboundary air pollution.

The specific undertakings of both Governments at this time are outlined below.

Interim Actions

1. Transboundary Air Pollution Agreement

Further to their Joint Statement of July 26, 1979, and subsequent bilateral discussion, both Governments shall take all necessary steps forthwith:

(a) to establish a Canada/United States Coordinating Committee which will undertake preparatory discussions immediately and commence formal negotiations no later than June 1, 1981, of a cooperative agreement on transboundary air pollution; and
(b) to provide the necessary resources for the Committee to carry out its work, including the working group structure as set forth in the Annex. Members will be appointed to the work groups by each Government as soon as possible.

2. Control Measures

To combat transboundary air pollution both Governments shall:

(a) develop domestic air pollution control policies and strategies, and as necessary and appropriate, seek legislative or other support to give effect to them;
(b) promote vigorous enforcement of existing laws and regulations as

they require limitation of emissions from new, substantially modified and existing facilities in a way which is responsive to the problems of transboundary air pollution; and
(c) share information and consult on actions being taken pursuant to (a) and (b) above.

3. Notification and Consultation

Both Governments shall continue and expand their long-standing practice of advance notification and consultation on proposed actions involving a significant risk or potential risk of causing or increasing transboundary air pollution, including:
(a) proposed major industrial development or other actions which may cause significant increases in transboundary air pollution; and
(b) proposed changes of policy, regulations or practices which may significantly affect transboundary air pollution.

4. Scientific Information, Research and Development

In order to improve understanding of their common problem and to increase their capability for controlling transboundary air pollution both Governments shall:
(a) exchange information generated in research programs being undertaken in both countries on the atmospheric aspects of the transport of air pollutants and on their effects on aquatic and terrestrial ecosystems and on human health and property;
(b) maintain and further develop a coordinated program for monitoring and evaluation of the impacts of transboundary air pollution, including the maintenance of a Canada/United States sampling network and exchange of data on current and projected emissions of major air pollutants; and
(c) continue to exchange information on research to develop improved technologies for reducing emissions of major air pollutants of concern.

The Memorandum of Intent will become effective on signature and will remain in effect until revised by mutual agreement.

DONE in duplicate at Washington, this fifth day of August, 1980, in the English and French languages, both texts being equally authoritative.

ANNEX

WORK GROUP STRUCTURE FOR NEGOTIATION OF TRANSBOUNDARY AIR POLLUTION AGREEMENT

I. Purpose

To establish technical and scientific work groups to assist in preparations for and the conduct of negotiations on a bilateral transboundary air pollution agreement. These groups shall include:
1. Impact Assessment Work Group
2. Atmospheric Modelling Work Group
3a. Strategies Development and Implementation Work Group
3b. Emissions, Costs and Engineering Assessment Subgroup
4. Legal, Institutional Arrangements and Drafting Work Group

II. Terms of Reference

A. *General*

1. The Work Groups shall function under the general direction and policy guidance of a Canada/United States Coordinating Committee co-chaired by the Department of External Affairs and the Department of State.
2. The Work Groups shall provide reports assembling and analyzing information and identifying measures as outlined in Part B below, which will provide the basis of proposals for inclusion in a transboundary air pollution agreement. These reports shall be provided by January 1982 and shall be based on available information.
3. Within one month of the establishment of the Work Groups, they shall submit to the Canada/United States Coordinating Committee a work plan to accomplish the specific tasks outlined in Part 8, below. Additionally, each Work Group shall submit an interim report by January 15, 1981.
4. During the course of negotiations and under the general direction and policy guidance of the Coordinating Committee, the Work Groups shall assist the Coordinating Committee as required.
5. Nothing in the foregoing shall preclude subsequent alteration of the tasks of the Work Groups or the establishment of additional Work Groups as may be agreed upon by the Governments.

B. *Specific*

The specific tasks of the Work Groups are set forth below.
1. Impact Assessment Work Group

Appendices 189

The Work Group will provide information on the current and projected impact of air pollutants on sensitive receptor areas, and prepare proposals for the "Research, Modelling and Monitoring" element of an agreement.

In carrying out this work, the Group will:
- identify and assess physical and biological consequences possibly related to transboundary air pollution;
- determine the present status of physical and biological indicators which characterize the ecological stability of each sensitive area identified;
- review available data bases to establish more accurately historic adverse environmental impacts;
- determine the current adverse environmental impact within identified sensitive areas—annual, seasonal and episodic;
- determine the release of residues potentially related to transboundary air pollution, including possible episodic release from snowpack melt in sensitive areas;
- assess the years remaining before significant ecological changes are sustained within identified sensitive areas;
- propose reductions in the air pollutant deposition rates—annual, seasonal and episodic—which would be necessary to protect identified sensitive areas; and
- prepare proposals for the "Research, Modelling and Monitoring" element of an agreement.

2. Atmospheric Modelling Work Group

The Group will provide information based on cooperative atmospheric modelling activities leading to an understanding of the transport of air pollutants between source regions and sensitive areas, and prepare proposals for the "Research, Modelling and Monitoring" element of an agreement. As a first priority the Group will by October 1, 1980, provide initial guidance on suitable atmospheric transport models to be used in preliminary assessment activities.

In carrying out its work, the Group will:
- identify source regions and applicable emission data bases;
- evaluate and select atmospheric transport models and data bases to be used;
- relate emissions from the source regions to loadings in each identified sensitive area;
- calculate emission reductions required from source regions to achieve proposed reductions in air pollutant concentration and deposition rates which would be necessary in order to protect sensitive areas;

- assess historic trends of emissions, ambient concentrations and atmospheric deposition trends to gain further insights into source receptor relationships for air quality, including deposition; and
- prepare proposals for the "Research, Modelling and Monitoring" element of an agreement.

3A. Strategies Development and Implementation Work Group

The Group will identify, assess and propose options for the "Control" element of an agreement. Subject to the overall direction of the Coordinating Committee, it will be responsible also for coordination of the activities of Work Groups I and II. It will have one subgroup.

In carrying out its work, the Group will:
- prepare various strategy packages for the Coordinating Committee designed to achieve proposed emission reductions;
- coordinate with other Work Groups to increase the effectiveness of these packages;
- identify monitoring requirements for the implementation of any tentatively agreed-upon emission-reduction strategy for each country;
- propose additional means to further coordinate the air quality programs of the two countries; and
- prepare proposals relating to the actions each Government would need to take to implement the various strategy options.

3B. Emissions, Costs and Engineering Assessment Subgroup

This Subgroup will provide support to the development of the "Control" element of an agreement. It will also prepare proposals for the "Applied Research and Development" element of an agreement.

In carrying out its work, the Subgroup will:
- identify control technologies, which are available presently or in the near future, and their associated costs;
- review available data bases in order to establish improved historical emission trends for defined source regions;
- determine current emission rates from defined source regions;
- project future emission rates from defined source regions for most probable economic growth and pollution control conditions;
- project future emission rates resulting from the implementation of proposed strategy packages, and associated costs of implementing the proposed strategy packages; and
- prepare proposals for the "Applied Research and Development" element of an agreement.

4. Legal, Institutional and Drafting Work Group

The Group will:
- develop the legal elements of an agreement such as notification

and consultation, equal access, non-discrimination, liability and compensation;
- propose institutional arrangements needed to give effect to an agreement and monitor its implementation; and
- review proposals of the Work Groups and refine language of draft provisions of an agreement.

APPENDIX D

Canada-Europe Ministerial Conference on Acid Rain
Ottawa, 21 March 1984

DECLARATION

The Governments of Austria, Canada, Denmark, Finland, France, the Federal Republic of Germany, the Netherlands, Norway, Sweden, and Switzerland, being Parties to the Convention on Long-Range Transboundary Air Pollution (hereinafter referred to as "the Convention"),

Determined to implement the principles and obligations regarding air pollution, including long-range transport of air pollutants, laid down in the Convention,

Recalling the decision of the United Nations Economic Commission for Europe (ECE) at its 38th session which stresses the urgency of intensifying efforts to arrive at coordinated national strategies and policies in the ECE region to reduce sulphur emissions effectively at national levels,

Recalling the recognition by the Executive Body of the Convention at its First Session of the need to decrease effectively the total annual emissions of sulphur compounds, or of their transboundary fluxes, by 1993–1995, using 1980 emissions levels as a basis for the calculation,

Noting the appreciation of the Executive Body of the Convention that a number of Contracting Parties are resolved to initiate measures for implementing a 30 percent reduction of national sulphur emissions or their transboundary fluxes by 1993–1995, using 1980 emission levels as a basis for the calculation of reductions,

Concerned that the present emissions of air pollution in Europe and North America are causing widespread damage to natural resources of vital importance, such as forests and waters, are damaging to materials and may have harmful health effects,

Recognizing the urgency of implementing reductions of annual sulphur emissions from those sources which make a significant contribution to the acidification of the environment,

Aware that the predominant sources of air pollution contributing to the acidification of the environment are the combustion of fossil fuels for energy production, industrial boilers and processes, individual house-heating and motor vehicles, which lead to emissions of sulphur dioxide and nitrogen oxides,

Convinced that air pollution abatement strategies for the reduction of emissions of sulphur dioxide, nitrogen oxides and other pollutants should be based on efficient measures such as energy saving and the application of the best available technologies which are economically feasible,

Recognizing that a reasonable time span is necessary for planning and implementing substantial reductions of emissions,

Aware that reducing emissions will have significant and positive results environmentally and economically, *Declare as follows:*

(1) The Signatories of this Declaration will implement reductions of national annual sulphur emissions by at least thirty percent as soon as possible and at the latest by 1993, using 1980 emission levels as the basis for the calculation of reductions;

(2) The Signatories recognize that a further reduction of sulphur emissions is or may prove necessary where environmental conditions warrant and should be considered as a matter of priority;

(3) The Signatories will, in their national policies and in international co-operation, take measures to decrease effectively the total annual emissions of nitrogen oxides from stationary and mobile sources as soon as possible and at the latest by 1993;

(4) The Signatories call upon other Parties to the Convention to join them, within the framework of the Convention, in implementing reductions of national annual sulphur emissions or of their transboundary fluxes by at least thirty percent by 1993, using 1980 emission levels as the basis for the calculation of reductions;

(5) The Signatories further stress the necessity of establishing within the framework of the Convention additional action for the purpose of achieving substantial reductions of emissions of other pollutants, especially nitrogen oxides.

NOTES

Chapter 1. The Emergence of the Acid Rain Controversy

1. "Acid rain" commonly refers to what is more precisely identified as the wet and dry processes for the deposition of acidic inputs to ecosystems (see U.S. Environmental Protection Agency 1980). Acidity is measured on the logarithmic pH scale, with pH being equal to the negative $\log_{10}$ of the hydrogen ion concentration. A solution that is neutral has a value of 7.0 on the pH scale. The "natural" acidity value often is assumed to be pH 5.6 calculated for distilled water in equilibrium with atmospheric carbon dioxide concentration. However, the presence of other naturally occurring substances—sulfur dioxide, ammonia, organic compounds, windblown dust—can produce "natural" values of pH 4.9 to 6.5 (Charlson and Rodhe 1982; Galloway et al. 1982).

2. A four-day fog in 1948 made almost 50 percent of the inhabitants of Donora, Pennsylvania—a small mill town dominated by steel and chemical plants—sick and twenty people died. Ten years later, residents who had been acutely ill during the fog had a higher rate of illness and died at an earlier age than the average for all Donora residents. Similar incidents occurred in the Meuse Valley of Belgium in the 1930s and the United Kingdom in the 1930s through the 1950s with a corresponding increase in the death rate.

3. Such estimates normally include voluntary pollution abatement expenditures, those necessitated by state and local regulation, and those necessitated by federal environmental regulation.

4. The "bubble" concept involves treating a firm's emissions from a plant as if the entire plant were covered by an imaginary bubble. This makes it possible to adjust the mix of control strategies within the plant. As a result, plant managers can vary emissions among different points in the plant in order to equalize marginal control costs among those sources. This produces lower total costs for the same aggregate level of emissions. EPA has used a bubble policy for air pollution control since December 1979.

196 NOTES TO PAGES 23–38

5. The use of offsets provides a means for allowing new sources that meet NSPS to be constructed in nonattainment areas. Construction is permitted if the new facility reduces the emissions of the same pollutant from existing sources already in compliance with air quality standards to a level more than equivalent to the prospective new plant's emissions.

6. The "good" versus "bad" science debate in the regulatory area involves both normative and empirical concerns. Empirically, some studies are much more uncertain than others in their findings and generalizability. Hence, they represent "bad" science and offer little help for rule-making from a value perspective.

Chapter 2. The Science of Acid Rain

1. Acidity of rain samples is calculated by chemical analyses that determine the equivalents of 3 anions and 5 cations. Bicarbonate (HCO_3^-) concentrations are calculated for samples where pH is above 5.0. For lower pH samples, the bicarbonate concentrations become insignificant and are not generally determined. Since the ion charges must be equal, the sums of anion equivalents must equal the sum of the cation equivalents. Any difference between the two sums is assumed to be hydrogen ion. The calculation is similar to the way profit is determined in simple bookkeeping. An example illustrates the procedure:

Anions ($\mu eq/l$)[a]		Cations ($\mu eq/l$)		
$SO_4^=$	72.5	Ca^{++}	38.4	
NO_3^-	21.1	Mg^{++}	15.6	
Cl^-	11.3	Na^+	20.9	
		K^+	3.6	
		NH_4^+	5.0	
		Subtotal	83.5	
		H^+	21.4	(by difference)
Total	104.9		104.9	

Note: For this example, calculated pH = 4.67 because 21.4 µeq/l = .0000214 eq/l, for which the logarithm is −4.67 (see Katzenstein 1983, 16).

a. µeq/l = microequivalents per liter.

2. These represent upper-bound estimates for natural emissions of sulfur oxides from biogenic, marine, and volcanic sources. Roth et al. (1985, 22) assert that the ratio of manmade to natural resources of NO_x emissions "could range 3:1 to 19:1, given uncertainties in estimates of natural emissions."

3. The Alkali Inspectorate was established by the Alkali Act of

Notes to Pages 41–86   197

1863 to regulate air pollution from the then emerging chemical industry in the United Kingdom. Currently called the Health and Safety Executive, it continues to be responsible for air pollution control.

4. The five net importers were Austria, Finland, Norway, Sweden, and Switzerland.

5. An isopleth map is created by connecting a series of points to form a contour line. All of the points are assumed to have a specified constant value for a given variable such as pH. This assumption makes it possible to make maps for large geographical areas with a limited number of observations or data points.

6. The acid neutralizing capacity (ANC) of lakes expressed in microequivalents per liter (µeq/l) of water provides a measure of their potential sensitivity to acidic inputs. The spectrum of relative sensitivity in the lake survey can be arrayed into intervals: acidified ANC ≤ 0 µeq/l; highly sensitive ≤ 50 µeq/l; slightly sensitive ≤ 200 µeq/l.

7. Leaching is a chemical process that separates soluble components by dissolving them out of soils by the action of percolating water.

8. The other panel members were: Richard Balzheiser, Electric Power Research Institute; Georgia Hidy, Environmental Research and Technology; Gene Likens, New York Botanical Garden; Stanford Penner, University of California, San Diego; Malvin Ruderman, Columbia University; Herman Postma, Oak Ridge National Laboratory; Michael Oppenheimer, Environmental Defense Fund; William Klemperer, Harvard University; and James Galloway, University of Virginia.

9. Substances such as calcium carbonate (limestone) can neutralize within limits the addition of acids to soils and water bodies in order to maintain desired hydrogen ion concentrations. As this buffering capacity increases, acidification decreases.

Chapter 3. Acid Rain Control Technologies and Mitigation Strategies

1. Errors in the data included in historical estimates and projections for emissions inventories may occur for a variety of reasons. Normally, true values for a specific geographic area are unknown. Estimated values are based on assumptions about the representativeness of emissions coefficients to source populations, fuel sulfur and nitrogen content, and compliance with regulations.

Chapter 4. The Economic Dimension

1. Economic analyses generally do no more than multiply natural science findings about service flow changes by an invariant price. The

researchers then speculate, if they recognize them at all, about how the resulting estimates would differ if price responses and agent adaptations were captured. Because of the differences in the behavior of emitters and receptors when a market in emissions rights does not exist, and because of the lack of parallel markets, economically efficient outcomes for the acid deposition problem may be impossible to trace exhaustively. The economic criterion is then reduced to whether those who gain from a change in precursor control could, in principle, compensate the losers and still have some residual gain. An excellent overview of the state of the art of benefits assessment is provided by Bentkover, Covello, and Mumpower (1986). For a discussion of the theoretical basis for applying the technique to acid deposition control, see Regens and Crocker (1984).

2. Personal communication, T. Crocker, November 1981. For an elaboration, see Crocker and Regens (1985).

3. Personal communication, P. Nyberg, Institute of Freshwater Research, Swedish National Board of Fisheries, November 1982.

4. Annualized costs equalize the yearly costs of a pollution control program over its lifetime. Total expenditures for each year are converted to present value expressed in constant dollars in order to adjust for the value of money and effect of inflation over time. These figures are then added and divided by the number of years projected for a program to provide an estimate of annualized costs.

5. A mill equals one-tenth of one cent.

6. Personal communication, T. Brand, Director of Environmental Affairs, Edison Electric Institute, March 1984.

7. The real rate is the cost after accounting for inflation. The real rate always equals the nominal rate minus the rate of inflation (see Kohler 1986).

8. With the exception of direct taxes imposed upon the states in proportion to population, the rule of liability to federal taxes must take no account of geography (see Florida v. Mellon, 273 U.S. 12 [1927]). While Congress is entitled to regulate by taxation (see Veazie Bank v. Fenno, 8 Wall. 533 [1869]; Mulford v. Smith 307 U.S. 38 [1939]), the uniformity clause may preclude applying constitutionally the revenue generation options examined in this section to some but not all states. Therefore, the appropriate base for examining these fees/taxes is probably all fifty states. For a more extended discussion of the uniformity clause, see Chase and Ducat (1974).

9. The British thermal unit (Btu) provides a measure for comparing the energy potential of different fuels. One Btu is equal to the amount of heat required to raise the temperature of one pound of

water one degree Fahrenheit under stated conditions of pressure and temperature. The Btu equivalents of common fuels are as follows:

Fuel (Common Measure)	Btu
Coal (1 ton)	24,000,000–28,000,000
Crude Oil (1 barrel)	5,800,000
Electricity (1 kilowatt hour)	3,412
Natural Gas (1 cubic foot)	1,032

Source: See Government Institutes (1977).

Chapter 6. Prospects for Policymaking

1. H.R. 2666 would reduce annual SO_2 emissions in two phases by 10 million tons and NO_x emissions by 4 million tons. Unlike H.R. 4567, this proposal would use monthly average rather than annual average rates to allocate SO_2 emissions.

2. The Executive Body is composed of national representatives of all countries that have ratified the ECE Convention and is responsible for developing protocols for implementing the terms of the convention.

3. States were grouped into the following regional categories: (1) East—Connecticut, Delaware, Maine, Maryland, Massachusetts, New Hampshire, New Jersey, New York, Pennsylvania, Rhode Island, Vermont; (2) Midwest—Illinois, Indiana, Iowa, Michigan, Minnesota, Missouri, Ohio, Wisconsin; (3) South—Alabama, Arkansas, Florida, Georgia, Kentucky, Louisiana, Mississippi, North Carolina, South Carolina, Tennessee, Virginia, West Virginia; and (4) West—Alaska, Arizona, California, Colorado, Hawaii, Idaho, Kansas, Montana, Nebraska, Nevada, New Mexico, North Dakota, Oklahoma, Oregon, South Dakota, Texas, Utah, Washington, Wyoming.

# REFERENCES

Abelson, P. H. 1985. "Technologies for Clean Use of Coal." *Science* 229 (August 30): 819.
Ackerman, B. A., and W. T. Hassler. 1981. *Clean Coal/Dirty Air*. New Haven, Conn.: Yale University Press.
Altshuller, A. P., and G. A. McBean. 1979. *The LRTAP Problem in North America: A Preliminary Overview Prepared by the United States—Canada Bilateral Research Consultation Group on the Long-Range Transport of Air Pollutants*. Downsview, Ontario, Canada: Atmospheric Environmental Service.
Ambe, Y., and M. Nishikawa. 1986. "Temporal Variation of Trace Element Concentrations in Selected Rainfall Events at Tsukuba, Japan." *Atmospheric Environment* 20 (October): 1931–40.
Ambio. 1976. "Report of the International Conference on the Effects of Acid Precipitation in Telemark, Norway." 5 (December): 200–01.
American Association for the Advancement of Science. 1986. *AAAS Report XI: Research and Development, FY 1987*. Washington, D.C.: AAAS.
Ashford, N. A. 1984. "The Use of Technical Information in Environmental, Health, and Safety Regulation: A Brief Guide to the Issues." *Science, Technology & Human Values* 9 (Winter): 130–33.
Atkinson, S. E. 1984. "The Effect of Global Optimization on Locally Optimal Pollution Control: Acid Rain." In *Economic Perspectives on Acid Deposition*, ed. T. D. Crocker, pp. 21–33. Woburn, Mass.: Butterworth Scientific Publishers.
Atkinson, S. E., and T. Tietenberg. 1982. "The Empirical Properties of Two Classes of Designs for Transferable Discharge Permit Markets." *Journal of Environmental Economics and Management* 9 (June):101–21.
Averch, H., and L. L. Johnson. 1962. "Behavior of the Firm Under Regulatory Constraint." *American Economic Review* 52 (December): 1052–69.

Barsin, J. A. 1984. "Options for Reducing NO_x, SO_2, During Combustion." In *The Acid Rain Sourcebook,* ed. T. C. Elliott and R. G. Schweiger, pp. 136–52. New York: McGraw-Hill.

Beamish, R. J., and H. H. Harvey. 1972. "Acidification of the LoChoche Mountain Lakes, Ontario, and Resulting Fish Mortalities." *Journal of the Fisheries Research Board of Canada* 29 (August): 1131–43.

Bengtsson, B., W. Dickson, and P. Nyberg. 1980. "Liming Acid Lakes in Sweden." *Ambio* 9 (January/February): 34–36.

Bentkover, J. D., V. T. Covello, and J. Mumpower. 1986. *Benefits Assessment: The State of the Art.* Boston: Reidel.

Berry, M. A. 1984. *A Method for Examining Policy Implementation: A Study of Decisionmaking for the National Ambient Air Quality Standards, 1964–1984.* EPA-600/X-84-091. Research Triangle Park, N.C.: U.S. Environmental Protection Agency.

Blodgett, L., L. Parker, and A. Backiel. 1984. *Acid Rain: Current Issues, Updated September 6.* Washington, D.C.: Congressional Research Service, Library of Congress.

Boyle, R. H., and R. A. Boyle. 1983. *Acid Rain.* New York: Shocken Books.

Braekke, F. H., ed. 1976. *Impact of Acid Precipitation on Forest and Freshwater Ecosystems in Norway.* Research Report 6/76. Oslo-As, Norway: SNSF Project.

Brenner, R. 1983. "Coal and Electricity Generation: An Economic Perspective." In *Costs of Coal Pollution Abatement,* ed. E. S. Rubin and I. M. Torrens, pp. 282–86. Paris: Organization for Economic Cooperation and Development.

Britt, D. L., and J. E. Fraser. 1983. "Effectiveness and Uncertainties Associated with the Chemical Neutralization of Acidified Surface Waters." In U.S. Environmental Protection Agency, *Proceedings of International Symposium on Lake Restoration, Protection and Management,* pp. 96–103. EPA 440/5-83-001. Washington, D.C.: EPA.

Bunsinger, J. A. 1986. "Evaluation of the Accuracy with Which Dry Deposition Can Be Measured with Current Micrometeorological Techniques." *Journal of Climatology and Applied Meteorology* 25 (August): 1100–24.

Burroughs, T. 1984. "Liberty Under Repair." *Technology Review* 87 (July): 63–69.

Canadian House of Commons. 1981. *Still Waters: The Chilling Reality of Acid Rain.* Report of the Sub-Committee on Acid Rain of the Standing Committee on Fisheries and Forestry. Ottawa, Canada: Canadian House of Commons.

Carroll, J. E. 1983. *Environmental Diplomacy.* Ann Arbor: University of Michigan Press.

Catalano, L., and J. Makansi. 1984. "Overview." In *The Acid Rain Sourcebook,* ed. T. C. Elliott and R. G. Schweiger, pp. 1–12. New York: McGraw-Hill.

Center for Policy Alternatives, Massachusetts Institute of Technology. 1980. *Benefits of Environmental Health and Safety Regulation.* Washington, D.C.: GPO.

Charlson, R. J., and H. Rodhe. 1982. "Factors Controlling the Acidity of Natural Rainwater." *Nature* 295 (February): 683–85.

Chartock, M. A., M. D. Devine, and E. M. Gunn. 1985. "Institutional Innovation for Cogeneration." *Policy Studies Review* 5 (August): 89–95.

Chase, H. S., and C. R. Ducat, eds. 1974. *Edwin S. Corwin's The Constitution and What It Means Today.* Princeton, N.J.: Princeton University Press.

Chubb, J. E. 1983. *Interest Groups and the Bureaucracy: The Politics of Energy.* Stanford, Calif.: Stanford University Press.

Claybrook, J. 1984. *Retreat from Safety: Reagan's Attack on America's Health.* New York: Pantheon Books.

Coase, R. H. 1960. "The Problem of Social Cost." *Journal of Law and Economics* 3 (October):1–44.

Cogbill, C. V., and G. E. Likens. 1974. "Acid Precipitation in the Northwestern United States." *Water Resources Research* 10 (December): 1133–37.

Conservation Foundation. 1984. *State of the Environment: An Assessment at Mid-Decade.* Washington, D.C.: Conservation Foundation.

Costle, D. 1981a. Letter from the EPA Administrator to the Honorable George Mitchell, U.S. Senator (January 15).

———. 1981b. Letter from the EPA Administrator to the Honorable Edmund Muskie, U.S. Secretary of State (January 23).

Council on Environmental Quality. 1977. *Environmental Quality.* Washington, D.C.: GPO.

———. 1979. *Environmental Quality.* Washington, D.C.: GPO.

———. 1980. *Environmental Quality.* Washington, D.C.: GPO.

———. 1981. *Environmental Quality.* Washington, D.C.: GPO.

———. 1985. *Environmental Quality.* Washington, D.C.: GPO.

Cowling, E. B. 1981. *An Historical Resume of Progress in Scientific and Public Understanding of Acid Precipitation and Its Consequences.* Oslo-As, Norway: SNSF Project.

———. 1982. "Acid Precipitation in Historical Perspective." *Environmental Science and Technology* 16 (February): 110A–123A.

Crandall, R. W. 1983a. *Controlling Industrial Pollution.* Washington, D.C.: Brookings.

———. 1983b. "Air Pollution, Environmentalists, and the Coal Lobby." In *The Political Economy of Deregulation: Interest Groups in the Regulatory Process,* ed. R. G. Noll and B. M. Owen, pp. 84–96. Washington, D.C.: American Enterprise Institute.

———. 1984. "An Acid Test for Congress." *Regulation* 8 (September/December): 21–28.

Crandall, R. W., and L. B. Lave, eds. 1981. *The Scientific Basis of Health and Safety Regulation.* Washington, D.C.: Brookings.

Crawford, M. 1985. "Utilities Look to New Coal Combustion Technology." *Science* 228 (May 3): 565.

Crocker, T. D. 1982. "Prior Information Required to Assess the Economic Benefits of Controlling Acid Deposition." In *Critical Assessment Document—Draft.* Washington, D.C.: EPA.

Crocker, T. D., and J. L. Regens. 1985. "Benefit-Cost Analyses of Acid Deposition Control: A Benefit-Cost Analysis: Its Prospects and Limits." *Environmental Science and Technology* 19 (February): 112–16.

Crocker, T. D., J. T. Tschirhart, and R. M. Adams. 1980. "A First Exercise in Assessing the Benefits of Controlling Acid Precipitation." In *Methods of Development for Assessing Acid Precipitation Control Benefits,* vol. 7. Washington, D.C.: EPA.

Daneke, G. A. 1982. "The Future of Environmental Protection: Reflections on the Difference Between Planning and Regulating." *Public Administration Review* 42 (May/June): 227–33.

Deutch, J., R. Balzhiser, G. Hidy, G. E. Likens, S. S. Penner, M. Ruderman, J. N. Galloway, W. Klemperer, M. Oppenheimer, and H. Postma. 1983. "Report of the Ad Hoc Committee to Review the National Acid Precipitation Assessment Program (NAPAP)." Prepared for the Science Advisory Board, EPA, Washington, D.C.

Devine, M. D., S. C. Ballard, I. L. White, M. A. Chartock, A. R. Brosz, F. J. Calzonetti, M. S. Eckert, T. A. Hall, R. L. Leonard, E. J. Malecki, G. D. Miller, E. B. Rappaport, and R. W. Rycroft. 1981. *Energy from the West: A Technology Assessment of Western Energy Resource Development.* Norman: University of Oklahoma Press.

Dickson, D. 1984. *The New Politics of Science.* New York: Pantheon Books.

———. 1986. "Europe Struggles to Control Pollution." *Science* 234 (December 12): 1315–16.

Downing, P. B. 1983. "Bargaining in Pollution Control." *Policy Studies Journal* 11 (June): 577–86.

Drablos, D., and A. Tollan, eds. 1980. *Ecological Impact of Acid Precipitation*. Oslo-As, Norway: SNSF Project.

Edison Electric Institute. 1983. *Statistical Yearbook of the Electric Utility Industry/1982*. Washington, D.C.: EEI.

Editorial Research Reports. 1982. *Environmental Issues: Prospects and Problems*. Washington, D.C.: Congressional Quarterly, Inc.

Electric Power Research Institute. 1983. "Developing the Options for Emissions Control." *EPRI Journal* 8 (November): 40–51.

Ember, L. 1985. "Rain Sulfates in West Linked to Sulfur Emissions." *Chemical and Engineering News* 63 (September 2): 18–19.

Energy Policy Project of the Ford Foundation. 1974. *A Time To Choose: America's Energy Future*. Cambridge, Mass.: Ballinger.

Energy Research and Advisory Board. 1985. *Clean Coal Use Technologies*. Washington, D.C.: U.S. Department of Energy.

Environment Canada. 1985. "Moving Ahead on Acid Rain." Ottawa: Environment Canada.

Fay, J. A., D. Golumb, and J. Gruhl. 1983. *Controlling Acid Rain*. Energy Laboratory Report No. MIT-EL83-004. Cambridge, Mass.: Energy Laboratory, MIT.

Federal/Provincial Research and Monitoring Coordinating Committee. 1986. *Assessment of the State of Knowledge on the Long-Range Transport of Air Pollutants and Acid Deposition*. Ottawa: Environment Canada.

Florida v. Mellon, 273 U.S. 12. 1927.

Fortune. 1983. "The Acid Rain Lobby Picks Up Steam." 107 (May 30), 33–36.

Frankel, E. 1986. "Technology, Politics and Ideology: The Vicissitudes of Federal Solar Energy Policy, 1974–1983." In *The Politics of Energy Research and Development,* ed. T. Byrne and D. Rich, pp. 67–87. New Brunswick, N.J.: Transaction Books.

Fraser, J. W., D. Hinckley, R. Burt, R. R. Severn, and J. Wisniewski. 1982. *A Feasibility Study to Utilize Liming as a Technique to Mitigate Surface Water Acidification*. EA-2362. Palo Alto, Calif.: Electric Power Research Institute.

Fraser, J. E., D. Minnick, H. Sverdrup, P. Warfvinge, and P. Saunders. 1984. *Economic Analysis of Historic Aquatic Base Addition Projects*. Reston, Va.: International Science and Technology, Inc.

Freeman, A. M. 1979. *The Benefits of Environmental Improvement: Theory and Practice*. Baltimore, Md.: Johns Hopkins University Press.

Freeman, G. C., Jr. 1983. "The Politics of Acid Rain." Presented at the fall meeting of the Board of Directors of the Southeastern Electric Exchange, Hilton Head, S.C.

———. 1985. "The U.S. Politics of Acid Rain." In *The Acid Rain Debate: Scientific, Economic, and Political Dimensions,* ed. E. J. Yanarella and R. H. Ihara, pp. 277–313. Boulder, Colo.: Westview Press.

Friedlaender, A. F., ed. 1978. *Approaches to Controlling Air Pollution.* Cambridge, Mass.: MIT Press.

Friedland, A. J., A. H. Johnson, and T. G. Siccama. 1986. "Zinc, Cu, Ni and Cd in the Forest Floor in the Northeastern United States." *Water, Air and Soil Pollution* 29 (July): 219–31.

Friends of the Earth. 1982. *Ronald Reagan and the American Environment.* San Francisco: Friends of the Earth.

Galloway, J. N., G. E. Likens, W. C. Keene, and J. M. Miller. 1982. "The Composition of Precipitation in Remote Areas of the World." *Journal of Geophysical Research* 87 (October): 8771–86.

Garvey, G. 1972. *Energy, Ecology, Economy: A Framework for Environmental Policy.* New York: Norton.

Gatti, J. F., ed. 1981. *The Limits of Government Regulation.* New York: Academic Press.

Giorgi, F. 1986. "A Partick Dry-Deposition Parameterization Scheme for Use in Tracer Transport Models." *Journal of Geophysical Research* 91 (August): 9797–9806.

Goldberg, V. P. 1976. "Regulation and Administration Contracts." *Bell Journal of Economics* 7 (Autumn): 426–48.

Goldfarb, R. S. 1980. "Compensating Victims of Policy Change." *Regulation* 4 (September/October): 22–30.

Gollop, F. M., and M. J. Roberts. 1985. "Cost-Minimizing Regulation of Sulfur Emissions: Regional Gains in Electric Power." *Review of Economics and Statistics* 67 (February): 81–90.

Gordon, R. L. 1984. "Problems of Modeling the Impacts of Acid Rain Legislation." In *Acid Rain Control: The Costs of Compliance,* ed. D. S. Gilleland and J. H. Swisher, pp. 11–50. Carbondale: Southern Illinois University Press.

Gould, R. 1985. *Going Sour: Science and Politics of Acid Rain.* Boston: Birkhauser.

Government Institutes. 1977. *Energy Reference Handbook.* Washington, D.C.: Government Institutes, Inc.

Green, L., Jr. 1984. "Coal Cleaning: The First Step," In *The Acid Rain Sourcebook,* ed. T. C. Elliott and R. G. Schweiger. New York: McGraw-Hill.

Hall, T. A. 1985. *Potential Natural Gas Substitution in Coal-Fired Power Plants: Draft Report.* Research Triangle Park, N.C.: Radian Corporation.

Hoffman, M. R. 1986. "On the Kinetics and Mechanism of Oxidation of Aquated Sulfur Dioxide by Ozone." *Atmospheric Environment* 20 (June):1145–54.

Hollander, E. 1983. "Federal Acid Rain Research," *EPRI Journal* 8 (November): 52–56.

Holt, N. A., and T. P. O'Shea. 1984. "General Status of IGCC Developments." In *The Acid Rain Sourcebook*, ed. T. C. Elliott and R. G. Schweiger. New York: McGraw-Hill.

Howard, R., and M. Perley. 1980. *Acid Rain: The Devastating Impact on North America*. New York: McGraw-Hill.

Huber, P. 1984. "The I-Ching of Acid Rain." *Regulation* 8 (September/December): 15–20; 56–65.

Hudson, H. E., Jr., and F. W. Gilcreas. 1976. "Health and Economic Aspects of Water Hardness and Corrosiveness." *Journal of the American Water Works Association* 68 (April): 201–04.

Hutton, C. A., and R. N. Gould. 1982. *Cleaning Up Coal: A Study of Coal Cleaning and the Use of Cleaned Coal*. Cambridge, Mass.: Ballinger.

ICF, Inc. 1983. *Analysis of Senate Emission Reduction Bill*. Washington, D.C.: ICF, Inc.

Imperato, P. J., and G. Mitchell. 1985. *Acceptable Risks*. New York: Viking/Penguin.

Indianapolis Star. 1984. "The Missing Pieces" (May 7).

Interagency Task Force on Acid Precipitation. 1985. *Annual Report, 1985: National Acid Precipitation Assessment Program*. Washington, D.C.: ITFAP.

International Energy Agency. 1985. *The Clean Use of Coal: A Technology Review*. Paris: IEA.

Jacobs, D. J. 1986. "Chemistry of OH in Remote Clouds and Its Role in the Production of Formic Acid and Peroxymonosulfate." *Journal of Geophysical Research* 91 (August): 9807–26.

James, B. R., and S. J. Riha. 1986. "pH Buffering in Forest Soil Organic Horizons: Relevance to Acid Precipitation." *Journal of Environmental Quality* 15 (July–September): 229–34.

Jones, C. O. 1975. *Clean Air*. Pittsburgh, Pa.: University of Pittsburgh Press.

Kahan, A. M. 1986. *Acid Rain: Reign of Controversy*. Golden, Colo.: Fulcrum.

Kash, D. E., and R. W. Rycroft. 1985. "Energy Policy: How Failure Was Snatched from the Jaws of Success." *Policy Studies Review* 4 (February): 433–44.

Kash, D. E., and R. W. Rycroft. 1984. *U.S. Energy Policy: Crisis and Complacency*. Norman: University of Oklahoma Press.

Katz, J. E. 1984. "The Uses of Scientific Evidence in Congressional

Policymaking: The Clinch River Breeder Reactor." *Science, Technology, and Human Values* 9 (Winter): 51–62.
Katzenstein, A. W. 1983. *Understanding Acid Rain*. Washington, D.C.: Edison Electric Institute.
———. 1986. "Acid Rain: A Further Look at the Evidence." *Power Engineering* 24 (March): 32–36.
Kingdon, J. W. 1984. *Agendas, Alternatives, and Public Politics*. Boston: Little, Brown.
Kneese, A. V. 1984. *Measuring The Benefits of Clean Air and Water*. Washington, D.C.: Resources for the Future.
Kohler, H. 1986. *Intermediate Microeconomics*. 2d ed. Glenview, Ill.: Scott, Foresman.
Kraft, M. E. 1982. "The Use of Risk Analysis in Federal Regulatory Agencies: An Exploration." *Policy Studies Review* 1 (May): 666–75.
Krug, E. C., and C. R. Frink. 1983. "Acid Rain on Acid Soil: A New Perspective." *Science* 221 (August 5):520–25.
Lash, J., K. Gillman, and D. Sheridan. 1984. *A Season of Spoils*. New York: Pergamon Books.
Lave, L. B., and G. S. Omenn. 1981. *Clearing the Air: Reforming the Clean Air Act*. Washington, D.C.: Brookings.
Lave, L. B., and E. P. Seskin. 1977. *Air Pollution and Human Health*. Baltimore, Md.: Johns Hopkins University Press.
Lefevre, S. R. 1985. "Some Lessons from Energy Concerning the Commercialization of New Technologies." *Policy Studies Review* 5 (August): 122–32.
Lefohn, A. S., and R. W. Brocksen. 1984. "Acid Rain Effects Research: A Status Report." *Journal of the Air Pollution Control Association* 34 (October): 1005–13.
Lewis, W. M., Jr., and M. C. Grant. 1980. "Acid Precipitation in the Western United States." *Science* 207 (January 11): 176–77.
Likens, G. E., F. H. Bormann, and N. M. Johnson. 1972. "Acid Rain." *Environment* 14 (March): 33–40.
Linthurst, R. A., D. H. Landers, J. M. Eilers, D. F. Brakke, W. S. Overton, E. P. Meir, and R. E. Crowe. 1986. *Characteristics of Lakes in the Eastern United States: Population Descriptions and Physico-Chemical Relations*. EPA/600-4-86/007a. Washington, D.C.: U.S. Environmental Protection Agency.
Liroff, R. A. 1986. *Reforming Air Pollution Regulation*. Washington, D.C.: Conservation Foundation.
Little, Arthur D., and Energy Ventures Analysis. 1984. *Economic Impacts of Alternative Acid Rain Control Strategies*. Cambridge, Mass.: Arthur D. Little, Inc.

Louma, J. R. 1980. *Troubled Skies, Troubled Waters: The Story of Acid Rain.* New York: Viking Press.

Lunt, R. R., and J. S. MacKenzie. 1984. "Use Regenerable Scrubbers to Slash SO_2 Emissions." In *The Acid Rain Sourcebook,* ed. T. C. Elliott and R. G. Schweiger, pp. 198–227. New York: McGraw-Hill.

MacAvoy, P. W. 1979. *The Regulated Industries and the Economy.* New York: Norton.

MacNeill, J. M. 1983. "Coal and Environment: Constraint or an Opportunity?" In *Costs of Coal Pollution Abatement,* ed. E. S. Rubin and I. M. Torrens, pp. 55–62. Paris: Organization for Economic Cooperation and Development.

Magat, W. A., ed. 1982. *Reform of Environmental Regulation.* Cambridge, Mass.: Ballinger.

Magnet, M. 1983. "How Acid Rain Might Dampen the Utilities." *Fortune* 108 (August 8): 57–64.

Mandelbaum, P. A., ed. 1985. *Acid Rain: Economic Assessment.* New York: Plenum.

Maraniss, D. 1984. "Congress' Search for an Acid Rainbow." *Washington Post National Weekly Edition* 1 (February 13): 6–7.

Marcus, A. 1980. "Environmental Protection Agency." In *The Politics of Regulation,* ed. J. Q. Wilson, pp. 267–303. New York: Basic Books.

Mares, J. W. 1983. "Testimony of Jan W. Mares, Assistant Secretary for Fossil Energy and Acting Director for Policy, Planning & Analysis, U.S. Department of Energy." In *Acid Rain: Implications for Fossil R&D,* hearings before the Subcommittee on Energy Development and Applications and the Subcommittee on Natural Resources, Agriculture Research and Environment, of the Committee on Science and Technology, U.S. House of Representatives, September 13, 20. Washington, D.C.: GPO.

Markowsky, J. J. 1984. "Pressurized Fluidized-Bed Combustion: A Future Option." In *The Acid Rain Sourcebook,* ed. T. C. Elliott and R. G. Schweiger, pp. 251–61. New York: McGraw-Hill.

Marshall, E. 1983. "Ruckelshaus Disappoints Canadians on Acid Rain." *Science* 222 (October 23): 401.

———. 1984. "Canada Goes It Alone on Acid Rain Controls." *Science* 223 (March 23): 1275.

———. 1987. "EPA Finds Western Lakes Free of Acid Pollution, But Vulnerable." *Science* 235 (January 23): 423.

Maulbetsch, J. S., M. W. McElroy, and D. Eskinazi. 1986. "Retrofit NO_x Control Options for Coal-Fired Electric Utility Power Plants." *Journal of the Air Pollution Control Association* 36 (November): 1294–98.

McCurdy, H. E. 1986. "Environmental Protection and the New Federalism: The Sagebrush Rebellion and Beyond." In *Controversies in Environmental Policy,* ed. S. Kamieniecki, R. O'Brien and M. Clarke, pp. 85–107. Albany: State University of New York Press.

McFarland, A. S. 1976. *Public Interest Lobbies: Decision Making on Energy.* Washington, D.C.: American Enterprise Institute.

McGlamery, G. G., and R. L. Torstick. 1976. "Cost Comparisons of Flue Gas Desulfurization Systems." In *Power Generation: Air Pollution Monitoring and Control,* ed. K. E. Noll and W. T. Davis. Ann Arbor, Mich.: Ann Arbor Science Publishers.

Meier, K. J. 1985. *Regulation: Politics, Bureaucracy and Economics.* New York: St. Martin's Press.

Melack, J. M., J. L. Stoddard, and C. A. Ochs. 1985. "Major Ion Chemistry and Sensitivity to Acid Precipitation of Sierra Nevada Lakes." *Water Resources Research* 21 (January): 27–32.

Melnick, R. S. 1984. "Pollution Deadlines and the Coalition for Failure." *Public Interest* 75 (Spring): 123–34.

Milbrath, L. W. 1984. *Environmentalists: Vanguard for a New Society.* Albany: State University of New York Press.

Mills, E. S. 1978. *The Economics of Environmental Quality.* New York: Norton.

Mosher, L. 1982. "Congress May Have to Resolve Stalled U.S.-Canadian Acid Rain Negotiations." *National Journal* 14 (March 13): 456–58.

———. 1983a. "Acid Rain Debate May Play a Role in the 1984 Presidential Sweepstakes." *National Journal* 15 (May 14): 1998–99.

———. 1983b. "Administration Loses Its Umbrella Against Steadfast Acid Rain Policy." *National Journal* 15 (July 30): 1590–91.

Mulford v. Smith, 307 U.S. 38. 1939.

National Commission on Air Quality. 1981. *To Breathe Clean Air.* Washington, D.C.: GPO.

National Research Council. 1981. *On Prevention of Significant Deterioration.* Washington, D.C.: National Academy Press.

———. 1983. *Acid Deposition Atmospheric Processes in Eastern North America: A Review of Current Scientific Understanding.* Washington, D.C.: National Academy Press.

———. 1986. *Acid Deposition: Long-Term Trends.* Washington, D.C.: National Academy Press.

Nierenberg, W. A., W. C. Ackermann, D. H. Evans, G. E. Likens, R. Patrick, K. A. Rahn, F. S. Rowland, M. A. Ruderman, and S. F. Singer. 1984. *Report of the Acid Rain Peer Review Panel.* Washington, D.C.: Office of Science and Technology Policy.

Noll, R. G., and B. M. Owen. 1983. "The Political Economy of Deregulation: An Overview." In *The Political Economy of Deregulation: Interest Groups in the Regulatory Process,* ed. Noll and Owen, pp. 26–52. Washington, D.C.: American Enterprise Institute.

North Atlantic Treaty Organization, Committee on Challenges of Modern Society. 1982. *Flue Gas Desulfurization Pilot Study Followup.* Brussels: NATO.

Oden, S. 1968. "The Acidification of Air and Precipitation and Its Consequences in the Natural Environment." *Ecology Committee Bulletin,* no. 1, Swedish National Science Research Council.

Office of Management and Budget. 1986. "Special Analysis K: Research and Development." *Budget of the United States Government, 1987.* Washington, D.C.: GPO.

Office of the Press Secretary, the White House. 1986. *Remarks by the President and Prime Minister Mulroney in Signing of NORAD Agreement and Statement on Acid Rain* (March 19). Washington, D.C.: Executive Office of the President.

Office of Science and Technology Policy. 1983. *Interim Report from OSTP's Acid Rain Peer Review Panel.* Washington, D.C.: Executive Office of the President.

Omernik, J. M., and G. E. Griffith. 1986. "Total Alkalinity of Surface Water: A Map of the Upper Midwest Region of the United States." *Environmental Management* 10 (November): 829–39.

Oppenheimer, M., C. B. Epstein, and R. E. Yuhnke. 1985. "Acid Deposition, Smelter Emissions, and the Linearity Issue in the Western United States." *Science* 229 (August 30): 859–62.

Organization for Economic Cooperation and Development. 1977. *The OECD Programme on Long-Range Transport of Air Pollutants.* Paris: OECD.

———. 1983. *Coal and Environmental Protection: Costs and Costing Methods.* Paris: OECD.

Ottar, B. 1976. "Monitoring Long-Range Transport of Air Pollutants: The OECD Study." *Ambio* 5 (December): 203–06.

Overrein, L. N., H. M. Seip, and A. Tollan. 1980. *Acid Precipitation—Effects on Forest and Fish.* Final Report of the SNSF Project, 1972–1980, SNSF Research Report no. 19. Oslo-As, Norway: SNSF Project.

Parker, L. B., and A. Kaufman. 1985. *Clean Coal Technology and Acid Rain Control: Birds of a Feather?* Congressional Research Service, Library of Congress (October 23). Washington, D.C.: CRS.

Parker, L. B., and R. E. Trumbule. 1983. *Mitigating Acid Rain with Technology: Avoiding the Scrubbing-Switching Dilemma.*

Congressional Research Service, Library of Congress (June 27). Washington, D.C.: CRS.

Pawlick, T. 1984. *A Killing Rain: The Global Threat of Acid Precipitation.* San Francisco: Sierra Club Books.

Poirot, R. L. 1986. "Visibility, Sulfate and Air Mass History Associated with the Summertime Aerosol in Northern Vermont." *Atmospheric Environment* 20 (July): 1457–69.

Polsby, N. W. 1984. *Political Innovation in America: The Politics of Policy Initiation.* New Haven, Conn.: Yale University Press.

Poundstone, W., and E. Rubin. 1985. *The Cleaning of Coal.* A Report to the Subcommittee on Energy Development and Applications of the Committee on Science and Technology. Washington, D.C.: GPO.

President's Commission for a National Agenda for the Eighties. 1980. *Energy, Natural Resources, and the Environment.* Washington, D.C.: GPO.

Quarles, J. 1976. *Cleaning Up America: An Insider's View of the Environmental Protection Agency.* Boston: Houghton Mifflin.

Ramo, S. 1981. "Regulation of Technological Activities: A New Approach." *Science* 215 (August 21): 837–42.

Regens, J. L. 1983. "The Regulatory Climate for Coal Development." In *Costs of Coal Pollution Abatement,* ed. E. S. Rubin and I. M. Torrens, pp. 104–11. Paris: Organization for Economic Cooperation and Development.

———. 1984. "Acid Rain: Does Science Dictate Policy or Policy Dictate Science?" In *Economic Perspectives on Acid Deposition,* ed. T. D. Crocker, pp. 5–19. Woburn, Mass.: Butterworth Scientific Publishers.

———. 1985a. "The Political Economy of Acid Rain." *Publius* 15 (Summer): 53–66.

———. 1985b. "Energy Research, Development and Demonstration: A Role for State Government?" *Environment* 27 (May): 19–20; 37–40.

Regens, J. L., and T. D. Crocker. 1984. "Applying Benefit-Cost Analysis to Acid Rain Control." *Management Science and Policy Analysis* 1 (Winter): 12–17.

Regens, J. L., T. M. Dietz, and R. W. Rycroft. 1983. "Risk Assessment in the Policy-Making Process." *Public Administration Review* 43 (March/April): 137–45.

Regens, J. L., and J. A. Donnan. 1986. "Uncertainty and Information Integration in Acidic Deposition Policymaking." *Environmental Professional* 8 (October): 342–50.

Regens, J. L., and R. W. Rycroft. 1985. "Perspectives on Acid Deposition: Science, Economics and Policy Making." In *The Acid Rain*

Debate, ed. E. J. Yanarella and R. H. Ihara, pp. 87–106. Boulder, Colo.: Westview.

Regens, J. L., and R. W. Rycroft. 1986. "Options for Financing Acid Rain Controls." *Natural Resources Journal* 26 (Summer): 519–49.

Reich, R. B. 1983. *The Next American Frontier.* New York: Times Books.

Rhoads, S. E. 1985. *The Economist's View of the World: Government, Markets, and Public Policy.* New York: Cambridge University Press.

Rhodes, S. L. 1984. "Superfunding Acid Rain Controls: Who Will Bear the Costs?" *Environment* 26 (July/August): 25–32.

Rhodes, S. L., and P. Middleton 1983. "The Complex Challenge of Controlling Acid Rain." *Environment* 25 (May):6–9; 31–37.

Rodhe, H., and M. J. Rood. 1986. "Temporal Evolution of Nitrogen Compounds in Swedish Precipitation since 1955." *Nature* 321 (June 19):762–64.

Roeder, P. W., and T. P. Johnson. 1985. "Public Opinion and the Environment: The Problem of Acid Rain." In *The Acid Rain Debate,* ed. E. J. Yanarella and R. H. Ihara, pp. 57–80. Boulder, Colo.: Westview.

Rosenbaum, W. A. 1985. *Environmental Politics and Policy.* Washington: Congressional Quarterly, Inc.

———. 1977. *The Politics of Environmental Concern.* 2d ed. New York: Praeger.

Roth, P., C. Blanchard, J. Harte, H. Michaels, M. T. El-Ashry. 1985. *The American West's Acid Test.* Washington, D.C.: World Resources Institute.

Rubin, E. S. 1983. "International Pollution Control Costs of Coal-Fired Power Plants." *Environmental Science and Technology* 17 (August): 366A–377A.

Rubin, E. S., and I. M. Torrens, eds. 1983. *Costs of Coal Pollution Abatement.* Paris: Organization for Economic Cooperation and Development.

Rushefsky, M. E. 1984. "The Misuse of Science in Governmental Policymaking." *Science, Technology and Human Values* 9 (Summer): 47–59.

Russell, C. S. 1982. "Externality, Conflict, and Decision." In *Regional Conflict and National Policy,* ed. K. A. Price, pp. 110–25. Baltimore, Md.: Johns Hopkins University Press.

Schmandt, J., and H. Roderick, eds. 1985. *Acid Rain and Friendly Neighbors.* Durham, N.C.: Duke University Press.

Scott, A. 1986. "The Canadian-American Problem of Acid Rain." *Natural Resources Journal* 26 (Spring): 337–58.

Seskin, E. P., R. J. Anderson, Jr., and R. O. Reid. 1983. "An Empirical Analysis of Economic Strategies for Controlling Air Pollution." *Journal of Environmental Economics and Management* 10 (June): 112–24.

Silverman, B. G. 1982. "Acid Rain and a Comparative Analysis of Two Large-Scale Electric Utility Models." *Journal of the Air Pollution Control Association* 32 (October): 1031–42.

Smith, R. A. 1972. *Air and Rain: The Beginnings of Chemical Climatology.* London: Longmans, Green.

Sonstelle, J. 1981. *Economic Incentives and the Revelation of Compliance Costs.* Report prepared for the Council on Environmental Quality, Washington, D.C.

Spencer, D. F., S. B. Alpert, and H. H. Gilman. 1986. "Cool Water: Demonstration of a Clean and Efficient New Coal Technology." *Science* 232 (May 2): 609–12.

Squires, A. M., M. Kwauk, and A. A. Avidan. 1985. "Fluid Beds: At Last, Challenging Two Entrenched Practices." *Science* 230 (December 20): 1329–37.

Stanfield, R. L. 1984. "Regional Tensions Complicate Search for an Acid Rain Remedy." *National Journal* 16 (May 5): 860–63.

———. 1985. "Environmentalists Try the Backdoor Approach to Tackling Acid Rain." *National Journal* 17 (October 19): 2365–68.

———. 1986a. "The Acid Rain Makers." *National Journal* 18 (June 14): 1500–03.

———. 1986b. "Summit Spurs Acid Rain Action." *National Journal* 18 (April 12): 892–93.

Stern, A. C., R. W. Boubel, D. B. Turner, and D. L. Fox. 1984. *Fundamentals of Air Pollution,* 2d ed. New York: Academic Press.

Stockman, D. 1980. Speech presented to the National Association of Manufacturers. Quoted in J. Claybrook, *Retreat from Safety,* p. 119. New York: Pantheon Books.

Sun, M. 1985. "Possible Acid Rain Woes in the West." *Science* 228 (April 5): 34–35.

———. 1986. "Acid Rain Plan Draws Mixed Review." *Science* 231 (January 24): 333.

Sununu, J. H. 1985. "Acid Rain: Sharing the Cost." *Issues in Science and Technology* 1 (Winter): 47–59.

Swedish Ministry for Foreign Affairs, Swedish Ministry of Agriculture. 1972. *Pollution Across National Boundaries: the Impact on the Environment of Sulfur in Air and Precipitation. Sweden's Case Study for the United Nations Conference on the Human Environment.* Stockholm: Royal Ministry for Foreign Affairs, Royal Ministry of Agriculture.

Swedish Ministry of Agriculture. 1982. *Acidification Today and Tomorrow: A Swedish Study Prepared for the 1982 Stockholm Conference on the Acidification of the Environment.* Stockholm: Royal Ministry of Agriculture.

Swedish National Environmental Protection Board. 1981. "Fösurning av mark och vatten." *Monitor,* Solna, Sweden.

Tearney, J. F., D. A. Froelich, and G. M. Graves. 1984. "Nonregenerable Wet FGD Controls SO_2 Emissions." In *The Acid Rain Sourcebook,* ed. T. C. Elliott and R. G. Schweiger, pp. 154–64. New York: McGraw-Hill.

Temple, Barker, and Sloane, Inc. 1985. *Economic Impact of Acid Rain Control Legislation: How Would U.S. Industry Be Affected?* Washington, D.C.: Edison Electric Institute.

Tobin, R. J. 1984. "Revising the Clean Air Act: Legislative Failure and Administrative Success." In *Environmental Policy in the 1980s: Reagan's New Agenda,* ed. N. J. Vig and M. E. Kraft, pp. 227–49. Washington, D.C.: Congressional Quarterly, Inc.

Tolchin, S. J., and M. Tolchin. 1983. *Dismantling America: The Rush to Deregulate.* Boston: Houghton Mifflin.

Torstrick, R. L., S. V. Tomlinson, J. R. Byrd, J. D. Veitch, and R. W. Gerstle. 1980. *Flue Gas Desulfurization Pilot Study.* EPA-EDT-114. Washington, D.C.: U.S. Environmental Protection Agency.

Ulrich, B. 1982. *Immissionsbelastungen Von Waldokosystemen (Dangers to the Forest Ecosystem Due to Acid Precipitation).* Special Communication from the Landesanstalt für Okologie Landschaftsentwicklung und Forstplanung Nordrhein-Westfalen. Göttingen, West Germany: Faculty of Forestry, University of Göttingen.

United Mine Workers. 1983. *Employment Impacts of Acid Rain.* Washington, D.C.: UMW.

United States v. Canada, 3 R. Intl'l Arb. Awards 1905. 1949.

U.S.-Canada Work Group 1. 1982. *Impact Assessment.* Final Report. Washington, D.C.: U.S. Environmental Protection Agency.

U.S.-Canada Work Group 2. 1982. *Atmospheric Science and Analysis.* Final Report. Washington, D.C.: U.S. Environmental Protection Agency.

U.S.-Canada Work Group 3B. 1982. *Emissions, Costs, and Engineering Assessment.* Final Report. Washington, D.C.: U.S. Environmental Protection Agency.

U.S. Department of Energy. 1983. *State Energy Data Report.* Washington, D.C.: DOE.

U.S. Department of Health, Education, and Welfare. 1969a. *The Cost of Clean Air: First Report of the Secretary of Health, Educa-*

tion and Welfare to the Congress of the United States in Compliance with Public Law 90-148. Washington, D.C.: GPO.

———. 1969b. *Progress Toward Cleaner Air.* Washington, D.C.: GPO.

U.S. Office of Technology Assessment. 1982. *The Regional Implications of Transported Air Pollutants: An Assessment of Acidic Deposition and Ozone.* Interim Draft. Washington, D.C.: OTA.

———. 1983. "An Analysis of the Sikorski/Waxman Acid Rain Control Proposal: H.R. 3400, 'The National Acid Deposition Control Act of 1983' " (July). Staff memorandum prepared by the OTA Oceans and Environment Program.

———. 1984. *Acid Rain and Transported Air Pollutants: Implications for Public Policy.* Washington, D.C.: GPO.

U.S. Congress. Senate. Committee on Energy and Natural Resources. 1982. *Acid Precipitation and the Use of Fossil Fuel.* Publication no. 97-87. Washington, D.C.: GPO.

———. Senate. Committee on Environment and Public Works. 1982. *Acid Rain: A Technical Inquiry.* Serial 97-H53. Washington, D.C.: GPO.

Veaze Bank v. Fenno, 8 Wall. 533. 1869.

Venkatram, A. 1986. "Statistical Long-Range Transport Models." *Atmospheric Environment* 20 (July): 1317–24.

Vermeulen, A. J. 1978. "Acid Precipitation in the Netherlands." *Environmental Science and Technology* 12 (September): 1016–21.

Vogelmann, H. M. 1982. "Catastrophe on Camel's Hump." *Natural History* 91 (November): 8–14.

Weaver, P. H. 1978. "Regulation, Social Policy, and Class Conflict." *Public Interest* 26 (Winter): 45–63.

Weidenbaum, M. L. 1977. *Business, Government, and the Public.* Englewood Cliffs, N.J.: Prentice-Hall.

Wetstone, G. S. 1983. "Paying for Acid Rain Control: An Introduction to the Trust Fund Approach." *Environmental Forum* 2 (August): 14–20.

Wetstone, G. S., and S. A. Foster. 1983. "Acid Precipitation: What Is It Doing to Our Forests?" *Environment* 25 (May): 10–12; 38–39.

Wetstone, G. S., and A. Rosencranz. 1983. *Acid Rain in Europe and North America.* Washington, D.C.: Environmental Law Institute.

Wilson, J. Q., ed. 1980. *The Politics of Regulation.* New York: Basic Books.

Wright, R. F., R. Harriman, A. Henriksen, B. Morrison, and L. A. Caines. 1980. "Acid Lakes and Streams in the Galloway Area, Southwestern Scotland." In *Ecological Impact of Acid Precipitation,* ed. D. Drablos and A. Tollan, pp. 248–49. Oslo-As, Norway: SNSF Project.

Yanarella, E. J. 1985. "The Foundations of Policy Immobilism Over Acid Rain Control." In *The Acid Rain Debate,* ed. E. J. Yanarella and R. H. Ihara, pp. 39–56. Boulder, Colo.: Westview.

Yeager, K. 1984. "Control Alternatives and the Acid Rain Issue." In *The Acid Rain Sourcebook,* ed. T. C. Elliott and R. G. Schweiger, pp. 82–97. New York: McGraw-Hill.

INDEX

Abelson, P. H., 80, 160
Acid deposition, definition of, 35. *See also* Acid rain
Acid Deposition Control Act of 1986, 153–56
Acid Precipitation Act of 1980, 8, 53–54, 118
"Acid Precipitation—Effects on Forest and Fish," 40
Acid rain. *See also* Air pollution; Control technologies; Economic issues; International issues; Nitrogen oxides; Political issues; Research and development
—causes of, 3–4, 36–37, 41–42, 53
—definition of, 35–37
—effects of, 5, 48–53; on agricultural crops, 50, 86, 87; on aquatic ecosystems, 48–86, 87, 88; and estimation of benefits of controls, 86–88, 156–57, 164; on forests, 48, 50, 86, 87, 88; on health, 51, 87, 163; international scope of, 5; on structures, 50–51, 86, 87
—geographic distribution of, 43–45
—monitoring of, 41–43
—natural vs. artificial sources of, 37, 45, 47
—schematic view of, 39
Acid Rain Control Act, 147
Ackerman, B. A., 17, 33, 115, 117, 130, 131
Acropolis, 50
Adams, R. M., 86
AFBC. *See* Atmospheric fluidized bed combustion
Air pollution: definition of, 178; health and welfare effects of, 6, 15; major sources of, 15. *See also* Acid rain

Air pollution control: and complexity of standard-setting, 12–17; delays in implementation of, 17–20; and economic efficiency, 20–23; and environmental effectiveness, 24–26; federal legislation re, 7–12, 53–54, 118; federal-state relationship re, 7, 9–10, 13, 17; of hazardous pollutants, 13, 16; history of, 5–7; market-oriented vs. regulatory approaches to, 21–25, 26, 28, 95–96; National Ambient Air Quality Standards for, 9, 10, 12–13, 14–15, 144; and social equity, 26–28. *See also* Clean Air Act; Control technologies; Cost allocation; Cost-benefit analysis; Political issues; Research and development
Air Pollution Control Act of 1955, 8
Air Quality Act of 1967, 7–9
Alkali Inspectorate (United Kingdom), 38
Altshuller, A. P., 149
Ambe, Y., 4
American Association for the Advancement of Science, 31, 81
Ammonia nitrogen, 53
Anions, 36
Aquatic ecosystems, 48, 86, 87, 88
Arsenic, 13
Asbestos, 13
Ash Council, 116
Ashford, N. A., 113–14
Atkinson, S. E., 95–96
Atmospheric fluidized bed combustion (AFBC), 76
Austria, 40, 150–51
Averch, H., 95

219

Backiel, A., 58
BACT. *See* Best available control technology
Balanced Budget and Emergency Deficit Control Act of 1985, 82
Barsin, J. A., 71
Beamish, R. J., 41
Beck, Robert A., 83
Becker, William S., 135
Belgium, 40, 151
Bengtsson, B. W., 80
Bennett, Kathleen M., 152
Benzene, 13
Berry, M. A., 11, 12, 14
Beryllium, 13
Best available control technology (BACT), 11, 13, 17
Bilateral Research Consultation Group on the Long-Range Transport of Air Pollutants, 144, 149
Blodgett, L., 58
Bormann, F. H., 41
Boyle, R. A., 156
Boyle, R. H., 127, 156
Brenner, R., 101
Britt, D. L., 80
Brocksen, R. W., 48
Btu tax, 104–06, 131
Bulgaria, 151
Bunsinger, J. A., 42
Burdick, Quentin, 156
Bureau of the Census, 22
Bureau of Economic Analysis, 22
Burford, Ann, 29, 120, 136
Burroughs, T., 50
Bush, George, 21
Byrd, Robert, 156

CAA. *See* Clean Air Act
Camel's Hump Mountain, 32, 129
Canada, 4, 30, 123, 125, 128, 152–53; bilateral agreements between U.S. states and provinces of, 143; Clean Air Act of, 144; creation of Acid Rain office by, 150; and EPA findings re international effects, 144, 146–47; Memorandum of Intent Between United States and, 43, 118, 149, 185–91; support of international emissions reductions by, 149–51. *See also* International issues
Canada-Europe Ministerial Conference on Acid Rain, 193–94
Canadian Coalition on Acid Rain, 125
Canadian National Film Board, 125
Canada/United States Coordinating Committee, 186, 188
Carbon monoxide (CO), 14
Carroll, J. E., 4
Carter, Jimmy, 118
Carter administration, 117, 119, 124
Catalano, L., 160
Cations, 36
CCC. *See* Chemical coal cleaning
Celeste, Richard, 141
CEM. *See* Continuous emissions monitoring
Center for Policy Alternatives, 123
CEQ. *See* Council on Environmental Quality
Charlson, R. J., 37
Chartock, M. A., 162
Chemical coal cleaning (CCC), 74–75
Chubb, J. E., 114, 121, 126
Citizen/Labor Coalition, 130
Claybrook, J., 31
Clean Air Act (CAA), 4, 118, 136, 140; action- and technology-forcing provisions of, 18–20, 59; economic efficiency of, 20–23; encouragement of political coalitions under, 34; federal-state relationship under, 7, 9–10, 13, 17; focus on local air quality under, 124; history of, 7, 8, 9–12; litigation under, 144, 146–47; public support of, 27; and Reagan administration, 32, 126; standard-setting under, 12–17; STAR program under, 144, 145; use of cost-benefit analysis under, 116. *See also* Air pollution control; Legislation, federal; New Source Performance Standards; State Implementation Plans
Clean Coal Technology Program, 81
Clinch River Breeder Reactor, 117
Coal: and Clean Coal Technology Program, 81; and economic effects of fuel-switching, 73, 91–94, 99–101; legislation re sulfur content of, 156; local use provisions re, 34; and New Source Performance Standards, 10–11, 13, 17, 89; as source

of precursor emissions, 118; sulfur content of, 93. *See also* Control technologies; Electric utilities; Sulfur dioxide
Coase, R. H., 106
Cogbill, C. V., 41
Coke ovens, 13
Connecticut, 146
Conservation Foundation, 3, 19
Construction-work-in-progress (CWIP), 98, 101
Conte, Silvio, 141
Continuous emissions monitoring (CEM), 107, 108
Control technologies, 59–61, 73–74, 160–61; costs of, 17, 20–22, 88–92, 94–96; dry flue gas desulfurization (FGD), 65, 67, 68; dual-register burners, 71; flue gas desulfurization (FGD) systems in U.S. utilities, 65; flue gas recirculation, 71; flue gas treatment, 73; indirect economic impact of, 91–94; low excess air (LEA), 71; low nitrogen oxide burners, 71–73; oil desulfurization, 63; physical coal cleaning (PCC), 61–63; policymaking implications of, 80–84; potential, 74–77; regenerable flue gas desulfurization (FGD), 69, 70; retrofitting proposals for, 130, 139; staged combustion, 69, 71; theoretical, 77–79; wet flue gas desulfurization (FGD), 63–64, 66. *See also* Air pollution control; Clean Air Act; Cost allocation; Flue gas desulfurization; Mitigation strategies
Convention on Long-Range Transboundary Air Pollution, 148, 150; text of, 177–84
Co-operative Programme for the Monitoring and Evaluation of the Long-Range Transmission of Air Pollutants in Europe (EMEP), 181–82. *See also* Convention on Long-Range Transboundary Air Pollution
Cost allocation: through Btu tax, 104–06, 131; distributional considerations in, 110–11, 132–34, 157; through electric utility ratemaking process, 97–102; through emissions fees, 105–06, 106–08; through federal budget outlays, 109–10, 131; through generation fees, 102–04, 105–06, 131–32; and market-oriented vs. regulatory approaches to air pollution control, 21–25, 26, 28, 95–96; "polluter-pays" principle for, 107, 117, 130–31, 148, 164; in proposed federal legislation, 138–41, 153–54; through sulfur-content-in-fuel tax, 108–09; under international law, 148. *See also* Cost-benefit analysis; Economic issues; Political issues

Cost-benefit analysis: and acid rain control costs, 88–92, 94–96; and control costs under Clean Air Act, 17–18, 20–23, 25; and estimation of environmental benefits of acid rain control, 85–88; and high costs vs. uncertain benefits of acid rain control, 5, 29–30, 96–97, 156–57, 163, 164; and meaning of "benefits," 122–23; social equity critique of, 26. *See also* Cost allocation; Economic issues; Political issues

Costle, Douglas, 119, 144, 146
Council of Economic Advisors, 21
Council on Environmental Quality (CEQ), 14, 16, 22, 47, 54, 119, 124, 147; on need for acid rain control, 118
Cowling, E. B., 38, 41
Crandall, R. W., 8, 9, 23, 97, 113, 114, 117, 157
Crawford, M., 77
Crocker, T. D., 48, 86, 88, 132, 161
CWIP. *See* Construction-work-in-progress
Czechoslovakia, 151

Daneke, G. A., 116
Davis, William, 60, 152
DDT, 32
Declaration on the Human Environment, 148. *See also* Stockholm Conference on the Human Environment
Democratic Republic of Germany, 151
Denmark, 40, 150–51

Department of Agriculture (USDA), 54
Department of Commerce, 22, 54
Department of Energy (DOE), 54; opposition to acid rain controls by, 120, 124, 126, 136; and research on acid rain control, 80, 81, 82, 160
Department of Health, Education, and Welfare (DHEW), 7, 9
Department of Health and Human Services (HHS), 54
Department of Justice, 7
Department of the Interior (DOI), 54, 126
Deutch, John, 55
Deutch panel, 55–56
Devine, M. D., 17
Dickson, D., 31, 119–20, 152
Dingell, John, 156
Donnan, J. A., 41, 45, 88, 157, 161, 162
Downing, P. B., 19–20, 24
Drablos, D., 40
Dry flue gas desulfurization (FGD), 65, 67, 68. *See also* Flue gas desulfurization
Dual-register burners, 71

EACN. *See* European Air Chemistry Network
Eastern Europe, 148
EBI. *See* Electron beam irradiation
Economic Commission for Europe (ECE), 148, 151, 193. *See also* Convention on Long-Range Transboundary Air Pollution
Economic issues: and economic efficiency arguments, 20–23; and environmental effectiveness, 24–26; and indirect economic impact of emissions controls, 73, 91–94, 99–101; and market-oriented vs. regulatory approaches to air pollution control, 21–25, 26, 28, 95–96; and need for more comprehensive economic analysis, 161. *See also* Cost allocation; Cost-benefit analysis
EDF. *See* Environmental Defense Fund
Edison Electric Institute (EEI), 124, 125
Editorial Research Reports, 19
EEI. *See* Edison Electric Institute

Electric Power Research Institute (EPRI), 63, 64, 73, 75, 124; funding of research by, 81; liming research by, 80
Electric utilities: compliance with Clean Air Act by, 130; control costs for, 91, 95; and cost allocation in proposed federal legislation, 153–54; flue gas desulfurization (FGD) systems in, 65; largest SO_2 emitters among, 139, 169–76; opposition to acid rain control by, 127, 137; rate-making process of, 97–102; as source of precursor emissions, 45, 47; sulfur content of coal produced for, 93. *See also* Coal; Control technologies; Cost allocation; Flue gas desulfurization
Electron beam irradiation (EBI), 77
Elemental sulfur, 69
Ember, L., 160
Emissions fees, 105–08. *See also* Cost allocation
Emissions reductions. *See* Control technologies; International issues; Legislation, federal; Legislation, state; Nitrogen oxides; Sulfur dioxide
Energy conservation, 78–79
Energy Policy Project, 82
Energy Research and Advisory Board, 75, 80
Energy Ventures Analysis, 94
Environmental Defense Fund (EDF), 122, 127, 159–60
Environmental Protection Agency (EPA), 9, 10–11, 13, 17, 54; bureaucratic strategy of, 115–17, 126; litigation against, 144, 146–47; research budget of, 31; State Acid Rain Program (STAR) of, 144, 145; sulfur dioxide emissions reductions proposal by, 136; survey of acid rain effects on lakes by, 44–45
Environment Canada, 150
EPRI. *See* Electric Power Research Institute
European Air Chemistry Network (EACN), 41, 53
European Economic Community (EEC), 152

Index 223

Fay, J. A., 91
Federal budget outlays, 109–10, 131. *See also* Cost allocation
Federal Republic of Germany, 40, 149, 150–51
FGD. *See* Flue gas desulfurization
FGT. *See* Flue gas treatment
Final Act of the Conference on Security and Co-operation in Europe, 177
Finland, 150–51
Flue gas desulfurization (FGD), 11, 45, 61, 82, 101; and Btu tax, 106; dry, 67, 68; economic impact of, 17; and generation fees, 104; vs. LIMB technology, 75, 160; and phased emissions reductions, 158; political controversy re, 28, 34, 73, 117; regenerable, 69, 70; retrofitting proposals for, 130, 139; and sulfur-content-in-fuel tax, 109; in U.S. utilities, 65; wet, 63–64, 66, 75. *See also* Control technologies
Flue gas recirculation, 71
Flue gas treatment (FGT), 73
Ford Foundation, 82
Fortune, 101–02
Foster, S. A., 50
France, 40, 150–51
Frankel, E., 79
Fraser, J. E., 80, 96
Freeman, G. C., 107, 139, 158, 161
Friedlaender, A. F., 9
Friedland, A. J., 50
Friends of the Earth, 32, 117, 123
Frink, C. R., 45
Froelich, D. A., 64
Fuel oil desulfurization, 63

Galloway, J. N., 37
Garvey, G., 59
Gatti, J. F., 21
Generation fees, 102–04, 105–06, 131–32. *See also* Cost allocation
Gilcreas, F. W., 80
Gillman, K., 27
Giorgi, F., 42
Glenn, John, 138
Goldfarb, R. S., 133
Gollop, F. M., 164
Gordon, Richard, 161
Gould, Roy, 61, 63, 113, 159
Gramm-Rudman-Hollings Act, 82

Grant, M. C., 43
Graves, G. M., 64
Green, L., Jr., 61
Griffith, G. E., 45

H.R. 3400, 121, 139
H.R. 4567, 153–56
Harvey, H. H., 41
Hassler, W. T., 17, 33, 115, 117, 130, 131
Hawkins, David G., 84
Heritage Foundation, 117
Hoffman, M. R., 42
Hollander, E., 32
Holt, N. A., 77
House Committee on Energy and Commerce, 138, 153, 156
House Subcommittee on Health and Environment, 153, 156
Howard, R., 110, 156
Hubbard Brook Experimental Forest, 43
Huber, Peter, 47, 59
Hudson, H. E., Jr., 80
Humphrey, Gordon, 141
Hutton, C. A., 63
Hydrocarbon (HC), 12, 53

IGCC. *See* Integrated gasification combined-cycle technology
Imperato, P. J., 21
Indiana, 143
Industrial Revolution, 6
Integrated gasification combined-cycle technology (IGCC), 76–77, 160
Interagency Task Force on Acid Precipitation, 8, 53–54, 57, 118–19, 136; report on causes of acid rain by, 120, 121, 124. *See also* National Acid Rain Precipitation Assessment Program
International Association of Machinists and Aerospace Workers, 117
International Energy Agency, 64
International issues, 4, 5, 30, 147–52; and administrative procedures under Clean Air Act, 10; and agreements between U.S. states and Canadian provinces, 143; and emissions reductions proposals, 148–52, 190, 194; and exchange of

information, 149, 180, 187; and international linkages to U.S. environmental coalition, 123, 125, 128; and litigation under Clean Air Act, 144, 146–47; and monitoring and evaluation, 181–82, 187; and "polluter-pays" principle, 148; and research and development, 179–80, 187, 189–90. *See also* Canada; Canada-Europe Ministerial Conference on Acid Rain; Convention on Long-range Transboundary Air Pollution; Memorandum of Intent Between the Government of Canada and the Government of the United States of America Concerning Transboundary Air Pollution
International Joint Commission, 144
Interstate pollution, 10, 144, 146
Ireland, 152
Italy, 40, 51

Jacobs, D. J., 42
James, B. R., 50
Johnson, A. H., 50
Johnson, L. L., 95
Johnson, N. M., 41
Johnson, T. P., 4
Jones, C. O., 9

Kahan, A. M., 127
Kash, D. E., 60
Katz, J. E., 113
Katzenstein, A. W., 48, 164
Kaufman, A., 63, 74
Kentucky, 143
Kingdon, J. W., 32
Kneese, A. V., 86
Kraft, M. E., 114
Krug, E. C., 45

Lash, J. K., 27, 31, 119
Lave, L. B., 6, 14, 23, 97, 113, 114, 157; on scientific defensibility of current standards, 12
LEA. *See* Low excess air
Lead (Pb), 14
League of Women Voters, 125
Lefevre, S. R., 83, 161
Lefohn, A. S., 48
Legislation, federal: history of, 7–12; proposed, 137–41, 153–56. *See also* Acid Precipitation Act of 1980; Clean Air Act; Political issues
Legislation, state, 141–44
Lewis, Drew, 60, 152
Lewis, W. M., 43
Lichtenstein, 151
Likens, G. E., 41
Limestone Injection Multistage Burner (LIMB), 54, 75, 94, 160
Liming, 80, 96, 158, 161
Linderman, Charles W., 84
Linthurst, R. A., 44
Liroff, R. A., 107
Litigation, 144, 146–47
Little, Arthur D., 94
Living Lakes, 80
Long-range transboundary air pollution, definition of, 178
Louma, J. R., 127
Love Canal, 32, 129
Low excess air (LEA), 71
Low nitrogen oxide burners, 71–73
Lugar, Richard, 137
Lunt, R. R., 69
Luxemborg, 151

MacAvoy, P. W., 21
McBean, G. A., 149
McCurdy, H. E., 17
McFarland, A. S., 122
McGlamery, G. G., 95
McGraw-Hill, Inc., 22
MacKenzie, J. S., 69
MacNeill, J. M., 148
Magat, W. A., 22
Magnet, M., 102
Magnetohydrodynamics (MHD), 77
Maine, 143, 144, 146
Makansi, J., 160
Mandelbaum, P. A., 85, 161
Maraniss, D., 140
Marcus, A., 24, 116
Mares, J. W., 82
Markowsky, J. J., 76
Marshall, E., 45, 150
Massachusetts, 142, 146
Maulbetsch, J. S., 71
Meier, R. J., 26
Melack, J. M., 45
Melnick, R. S., 19
Memorandum of Intent (MOI) Between the Government of Canada and the Government of the United

States of America Concerning Transboundary Air Pollution, 43, 118, 149; text of, 185–91
Mercury, 13
Methodist church, 117
Michigan, 143
Middleton, P., 110–11, 157
Milbrath, L. W., 123
Mills, E. S., 107
Mitchell, G., 21
Mitchell, George, 137, 144, 156
Mitigation strategies, 80, 96, 158, 161. *See also* Control technologies
MOI. *See* Memorandum of Intent
Mosher, L., 127, 130, 138, 152, 165
Motor Vehicle Pollution Control Act of 1965, 7
Mulroney, Brian, 149, 153
Muskie, Edmund, 144

NAAQS. *See* National Ambient Air Quality Standards
NAPAP. *See* National Acid Precipitation Assessment Program
National Academy of Sciences (NAS), 51, 120, 121, 124
National Acid Precipitation Assessment Program (NAPAP), 8, 53–58, 80. *See also* Interagency Task Force on Acid Precipitation
National Aeronautics and Space Administration (NASA), 54
National Ambient Air Quality Standards (NAAQS), 10, 12–13, 14–15, 144
National Coal Association, 125
National Commission on Air Quality (NCAQ), 25
National emissions standards for hazardous air pollutants (NESHAPS), 13
National Environmental Policy Act (NEPA), 9, 116
National Governors Association (NGA), 141
National Oceanic and Atmospheric Administration (NOAA), 54
National Research Council, 17, 42, 44, 47, 51, 164
National Science Foundation, 54
National Taxpayer's Union, 117
National Wildlife Federation (NWF), 122, 125, 146

Natural Resources Defense Council (NRDC), 117, 122, 127, 146
NESHAPS. *See* National emissions standards for hazardous air pollutants
Netherlands, 40, 150–51
New Clean Air Act, 138
"New federalism," 17
New Hampshire, 142, 146
New Jersey, 146
New Mexico Public Service, 69
New Source Performance Standards (NSPS), 9, 10–11, 13; economic impact of, 17, 22; effectiveness of, 89–90
New York, 144, 146; Acid Deposition Control Act, 142, 143
New York v. Thomas, 146–47
Nierenberg, W. A., 120, 136
Nishikawa, M., 4
Nitrates, 42, 53, 57
Nitrogen oxides, (NO_x), 4, 94, 118; available control technologies for, 69–74; as cause of acid rain, 42, 53; geographic distribution of emissions of, in U.S. and Canada, 46; National Ambient Air Quality Standard for, 14–15; potential control technologies for, 54, 75–77; proposals for reductions in emissions of, 91, 151, 158; sources of emissions of, 37, 45, 47; state efforts to control, 143; top 100 plants emitting, 169–76; trends in emissions of, 79, 89. *See also* Control technologies; Sulfur dioxide
Nixon, Richard, 9, 116
Nixon administration, 115, 116
NOAA. *See* National Oceanic and Atmospheric Administration
Noll, R. G., 122
North Atlantic Treaty Organization (NATO), 63
Northern Illinois Public Service Company, 69
Norway, 40, 150–51; Institute for Air Research, 41; Interdisciplinary Research Program (SNSF Project), 40
NRDC. *See* Natural Resources Defense Council
NSPS. *See* New Source Performance Standards
Nuclear power, 78

Obrentz, Margery, 101
Oden, Svante, 38
Office of Management and Budget (OMB), 81, 120, 136
Office of Science and Technology Policy (OSTP), 63, 120, 121, 124, 136
Office of Technology Assessment (OTA), 50, 58, 61, 71, 75, 79, 92, 94, 95, 139, 157; on delays in acid rain control, 57
Oil desulfurization, 63
Omenn, G. S., 12, 14, 23
Omernik, J. M., 44
Ontario, 143
Oppenheimer, M., 159
Organization for Economic Cooperation and Development (OECD), 40–41, 51, 148
O'Shea, T. P., 77
OSTP. *See* Office of Science and Technology Policy
Ottar, B., 41
Overrein, L. N., 40
Owen, B. M., 122
Ozone (O_3), 12, 14, 53

Parker, L., 58, 61, 63, 64, 71, 74, 75, 77, 161
Particulate Matter (PM), 14
Pawlick, T., 121
PCC. *See* Physical coal cleaning
Pennsylvania, 144
Perhac, Ralph, 124
Perley, M., 110, 156
PFBC. *See* Pressurized fluidized bed combustion
Physical coal cleaning (PCC), 61–63
Poirot, R. L., 51
Political issues: and acid rain as surrogate issue, 4–5, 28–29, 33–34, 117–18, 157; and bureaucratic politics, 114–17; during Carter administration, 118–19, 124; and client politics, 126–29; and coalition opposing emissions controls, 34, 124, 127–28; and coalition supporting emissions controls, 34, 117, 122–23, 125, 128–29, 157; and distributional consequences of policy choices, 110–11, 132–34, 157; and entrepreneurial politics, 32–33, 129–31; and environmental effectiveness, 24–26; and high costs vs. uncertain benefits of acid rain control, 5, 29–30, 96–97, 156–57, 163, 164; and interest-group initiatives, 121–26; and linkages between science and policy, 51–53, 113–14, 135–36, 156–57; and majoritarian politics, 131–32; and phased emissions reductions proposals, 157–58, 161; and policy dilemmas, 162–65; and "polluter-pays" principle, 107, 117, 130–31, 148, 164; and proposals for hybrid programs, 162; and public awareness of acid rain issue, 4, 32–33; and R&D policy as delaying tactic, 30–31, 119–20, 127, 133, 136–37; and regional disputes, 124, 130, 139–41, 154–56; and social equity, 26–28. *See also* Air pollution control; Clean Air Act; Cost allocation; Economic issues; Legislation, federal; Legislation, state; Reagan administration
Polsby, N. W., 32
Potential hydrogen (ph) scale, 35–36
Poundstone, W., 75
President's Commission for a National Agenda for the Eighties, 25
Pressurized fluidized bed combustion (PFBC), 76, 160
Prevention of significant deterioration (PSD), 10, 13, 17, 22
Provincial Research and Monitoring Coordinating Committee, 43
Proxmire, William, 141
PSD. *See* Prevention of significant deterioration
Public Health Services, 7

Quarles, J., 116

Radionuclides, 13
Ramo, S., 114
Reagan, Ronald, 119, 124, 136, 149, 152, 153
Reagan administration: and acid rain R&D, 30–33, 81, 119–20, 127, 136–37; deregulation policies of, 27, 29, 119, 133; international isolation of, 152; opposition to emissions controls by, 136–37, 153, 154; polarization of acid rain issue

Index 227

by, 125–26; and privatization of technology development, 81, 161
Regenerable flue gas desulfurization (FGD), 69, 70f. *See also* Flue gas desulfurization
Regens, James L., 5, 6, 41, 42, 43, 45, 48, 78, 86, 88, 91, 113, 132, 157, 161; on acid rain policy dilemmas, 162; on NSPS effectiveness, 89–90; on timing of acid rain regulation, 164–65
Reich, R. B., 133–34
Research and development: in Canada, 150; on causes and effects of acid rain, 38–41, 86–88, 120, 159–60; re control technologies, 80–84, 160–61; federal funding of, 31, 54, 55, 56, 81–82; importance of, 159–60, 163; international cooperation in, 179–80, 187, 189–90; as policy focus, 30–33, 119–20, 127, 136–37; scope of, 31–32, 54, 57–58; state legislation re, 142–43; task forces for, 53–54; timing of, 54–57. *See also* Acid rain, effects of; Control technologies
Rhoads, S. E., 163
Rhode Island, 144
Rhodes, S. L., 110–11, 132, 157
Riha, S. J., 50
Roberts, M. J., 164
Roderick, H., 4
Rodhe, H., 37, 53
Roeder, P. W., 4
Rood, M. J., 53
Rosenbaum, W. A., 11, 117, 125, 132
Rosencranz, A., 149
Roth, P., 159
Rothchild, Edwin, 130
Rubin, E. S., 22, 75, 95
Ruckelshaus, William, 21, 55, 116, 120, 136
Rural America, 117
Rushefsky, M. E., 31, 113
Russell, C. S., 124, 133
Rycroft, R. W., 5, 60

S. 2203, 13
Scalia, Antonin, 146
Schmandt, J., 4
Scotland, 80
SCR. *See* Selective catalytic reduction

Scrubbers. *See* Flue gas desulfurization
Seip, H. M., 40
Selective catalytic reduction (SCR), 73
Selective noncatalytic reduction (SNR), 73
Senate Committee on Environment and Public Works, 138, 156
Senate Subcommittee on Environmental Protection, 156
Seskin, E. P., 6, 95
Sheridan, D., 27
Siccama, T. G., 50
Sierra Club, 122, 146
Sierra Club v. Ruckelshaus, 147
Sikorski, Gerry, 121, 139, 153–54
Silverman, B. G., 95
SIP. *See* State implementation plan
Smith, Robert Angus, 38
SNR. *See* Selective noncatalytic reduction
SNSF Project, 40
Solar energy, 78
Sonstelie, J., 22
Southern Governors' Association, 140
Soviet Union, 151
Spain, 152
Spencer, D. F., 77
Stafford, Robert, 138, 147, 156
Staged combustion, 69, 71
Stanfield, R. L., 84, 111, 135, 152, 153, 158, 162
STAPPA. *See* State and Territorial Air Pollution Program Administrators
State Acid Rain Program (STAR), 144, 145
State and Territorial Air Pollution Program Administrators (STAPPA), 135, 144
State implementation plan (SIP), 10, 13, 95, 144, 146
Statue of Liberty, 50
Stern, A. C., 7
Stockholm Conference on the Human Environment: 1972, 38, 40, 118, 147, 148, 177; 1982, 150, 151
Stockman, David, 85, 120
Sulfates, 42, 53, 57
Sulfur, sources of emissions of, 37
Sulfur byproducts, 69

Sulfur-content-in-fuel tax, 108–09
Sulfur dioxide (SO$_2$), 4; as cause of acid rain, 41–42, 48, 53; costs of reductions in emissions of, 92, 94–96, 164; current emissions standards for, 118; and economic effects of fuel-switching, 91–94, 99–101; geographic distribution of emissions of, in U.S. and Canada, 46; international proposals for reductions in emissions of, 148–52; legislative proposals for reductions in emissions of, 53, 137–42, 158; National Ambient Air Quality Standard re, 14–15; New Source Performance Standards for, 11, 17; potential control technologies for, 54, 74–77; scientific basis for control of emissions of, 51–53; sources of emissions of, 37, 45, 47; top 100 plants emitting, 169–76; transboundary deposition of, 41; trends in emissions of, 79, 89. *See also* Control technologies; Cost allocation; Nitrogen oxides
Sulfuric acids, 4, 69
Sun, M., 60, 160
Sununu, J. H., 141, 162
Superfund for hazardous wastes, 102
Sweden, 40, 123, 149, 150–51; liming in, 80, 96; Ministry of Agriculture, 3, 40, 49, 149; Ministry for Foreign Affairs, 40; National Environmental Protection Board, 35
Switzerland, 40, 151

Tall stacks, 45, 47
Tearney, J. F., 64
Temple, Barker, and Sloane, Inc., 94
Tennessee Valley Authority (TVA), 54
"30 Percent Club," 151
Three Mile Island, 129
Tietenberg, T., 95
Tobin, R. J., 29, 119, 125–26
Tolchin, M., 126
Tolchin, S. J., 126
Tollan, A., 40
Torrens, I. M., 22
Torstrick, R. L., 63, 95
Toxic Air Pollutant Evaluation and Control Program, 13, 16
Trail Smelter Case, 148
Trudeau, Pierre Elliott, 149

Trumbule, R. E., 61, 63, 64, 71, 75, 77, 161
Tschirhart, J. T., 86

Ulrich, B., 41
Union of Concerned Scientists, 117
United Kingdom, 40, 151, 152
United Mine Workers, 121, 124
United Nations: Charter of, 148, 177; Economic Commission for Europe (ECE), 148, 193; 1972 Conference on the Human Environment, 38, 40, 118, 147, 148, 177; 1982 Conference on the Human Environment, 150, 151
United States v. Canada, 148
U.S.-Canada Memorandum of Intent on Transboundary Air Pollution (MOI), 43, 118, 149; text of, 185–91
U.S.-Canada Research Consultation Group on Long-Range Transboundary Air Pollution, 144, 149
U.S.-Canada Work Groups, 120; Group 1, 80, 88; Group 3B, 61, 63, 67, 69, 74, 89, 91, 95, 138

Venkatram, A., 47
Vermeulen, A. J., 41
Vermont, 144, 146
Vinyl chloride, 13
VOC. *See* Volatile organic compound
Vogelmann, H. M., 50
Volatile organic compound (VOC), 89

Water pollution, 27
Watt, James, 29
Waxman, Henry, 138–39, 153–54
Waxman-Sikorski Bill, 121, 139
Weaver, P. H., 123
Weidenbaum, Murray, 21
Wellman-Lord Regenerable FGD Process, 69, 70f.
Wet flue gas desulfurization (FGD), 63–64, 66f., 75. *See also* Flue gas desulfurization
Wetstone, G. S., 50, 102, 149
Wilson, J. Q., 32, 126–27, 129, 131
Winters, Dwaine, 160
Wisconsin, 142
World Resources Institute, 159
Wright, R. F., 41

Yanarella, Ernest, 135
Yeager, K., 61

Pitt Series in Policy and Institutional Studies
Bert A. Rockman, Editor

The Acid Rain Controversy
James L. Regens and Robert W. Rycroft

Agency Merger and Bureaucratic Redesign
Karen M. Hult

The Aging: A Guide to Public Policy
Bennett M. Rich and Martha Baum

Clean Air: The Policies and Politics of Pollution Control
Charles O. Jones

Comparative Social Systems: Essays on Politics and Economics
Carmelo Mesa-Lago and Carl Beck, Editors

Congress and Economic Policymaking
Darrell M. West

Congress Oversees the Bureaucracy: Studies in Legislative Supervision
Morris S. Ogul

Foreign Policy Motivation: A General Theory and a Case Study
Richard W. Cottam

Homeward Bound: Explaining Changes in Congressional Behavior
Glenn Parker

Imagery and Iedology in U.S. Policy Toward Libya, 1969–1982
Mahmoud G. ElWarfally

The Impact of Policy Analysis
James M. Rogers

Iran and the United States: A Cold War Case Study
Richard W. Cottam

Japanese Prefectures and Policymaking
Steven R. Reed

Managing the Presidency: Carter, Reagan, and the Search for Executive Harmony
Colin Campbell, S.J.

Organizing Governance, Governing Organizations
Colin Campbell, S.J., and B. Guy Peters, Editors

Perceptions and Behavior in Soviet Foreign Policy
Richard K. Herrmann

Pesticides and Politics: The Life Cycle of a Public Issue
Christopher J. Bosso

Policy Analysis by Design
Davis B. Bobrow and John S. Dryzek

Political Leadership: A Source Book
Barbara Kellerman

The Politics of Public Utility Regulation
William T. Gormley, Jr.

The Politics of the U.S. Cabinet: Representation in the Executive Branch, 1789–1984
Jeffrey E. Cohen

The Presidency and Public Policy Making
George C. Edwards III, Steven A. Shull, and Norman C. Thomas, Editors

Public Policy in Latin America: A Comparative Survey
John W. Sloan

Roads to Reason: Transportation, Administration, and Rationality in Colombia
Richard E. Hartwig

The Struggle for Social Security, 1900–1935
Roy Lubove

The U.S. Experiment in Social Medicine: The Community Health Center Program, 1965–1986
Alice Sardell